Hybrid Economic-Environmental Accounts

National Accounting Matrices of Environmental Accounts (NAMEA) tables are used to analyze a range of environmental pressures and economic data resulting from consumption and production patterns – helping us gain a far better notion of the consequences of individuals', households' and firms' actions for the world we live in. This book deals with the increasingly complex issues of hybrid environmental and economic accounts. The perspective of environmental accounting for the analysis of the relationships between the economic and environmental systems, especially regarding the satellite accounts like NAMEA, is relatively recent, and partly derives from the conceptual and applied deficits that have emerged during the setting up of green GDP or GNP measures as alternative measures of accounting.

NAMEA provides a comprehensive and integrated picture of the economic system in association with the environmental system (physical pressures such as emissions) by a sector classification. This book is an integrated collection of complementary papers that revolve around the issue of environment-economic accounting. In the first part a historical background and empirical issues related to the NAMEA-type table definitions and estimations open the book, followed by some applications and analyses mainly applied to a sub-national level. The second part opens the window to international case studies for different EU countries and studies with methodological insights.

These policy-oriented, original works are primarily from an applied perspective, although theoretical aspects are also fully developed. The book should be of use to Environmental and Ecological economics students and researchers, as well as those studying the more general field of Environment studies.

Valeria Costantini is currently lecturer in Environmental Economics and Urban Economics and assistant professor at the University of Roma Tre, Italy.

Massimiliano Mazzanti is currently lecturer in Environmental Economics and associate professor at the University of Ferrara, Italy.

Anna Montini is assistant professor in Economics and lecturer in Economics and Environmental Economics at the University of Bologna, Italy.

Routledge studies in ecological economics

1 **Sustainability Networks**
Cognitive tools for expert collaboration in social-ecological systems
Janne Hukkinen

2 **Drivers of Environmental Change in Uplands**
Aletta Bonn, Tim Allot, Klaus Hubaceck and Jon Stewart

3 **Resilience, Reciprocity and Ecological Economics**
Northwest coast sustainability
Ronald L. Trosper

4 **Environment and Employment**
A reconciliation
Philip Lawn

5 **Philosophical Basics of Ecology and Economy**
Malte Faber and Reiner Manstetten

6 **Carbon Responsibility and Embodied Emissions**
Theory and measurement
João F.D. Rodrigues, Alexandra P.S. Marques and Tiago M.D. Domingos

7 **Environmental Social Accounting Matrices**
Theory and applications
Pablo Martínez de Anguita and John E. Wagner

8 **Greening the Economy**
Integrating economics and ecology to make effective change
Bob Williams

9 **Sustainable Development**
Capabilities, needs, and well-being
Edited by Felix Rauschmayer, Ines Omann and Johannes Frühmann

10 The Planet in 2050
The Lund discourse of the future
Edited by Jill Jäger and Sarah Cornell

11 Bioeconomics
Edited by Mauro Bonaiuti

12 Socioeconomic and Environmental Impacts on Agriculture in the New Europe
Post-Communist transition and accession to the European Union
S. Serban Scrieciu

13 Waste and Recycling
Theory and Empirics
Takayoshi Shinkuma and Shunsuke Managi

14 Global Ecology and Unequal Exchange
Fetishism in a zero-sum world
Alf Hornborg

15 The Metabolic Pattern of Societies
Where economists fall short
Mario Giampietro, Kozo Mayumi and Alevgül H. Sorman

16 Energy Security for the EU in the 21st Century
Markets, geopolitics and corridors
Edited by José María Marín-Quemada, Javier García-Verdugo and Gonzalo Escribano

17 Hybrid Economic-Environmental Accounts
Edited by Valeria Costantini, Massimiliano Mazzanti and Anna Montini

Hybrid Economic-Environmental Accounts

Edited by Valeria Costantini, Massimiliano Mazzanti and Anna Montini

Routledge
Taylor & Francis Group
LONDON AND NEW YORK

First published 2012
by Routledge
2 Park Square, Milton Park, Abingdon, Oxfordshire OX14 4RN
Simultaneously published in the USA and Canada
by Routledge
711 Third Avenue, New York, NY 10017
First issued in paperback 2014
Routledge is an imprint of the Taylor & Francis Group, an informa business

British Library Cataloguing in Publication Data
A catalogue record for this book is available from the British Library

Library of Congress Cataloging in Publication Data
Hybrid economic-environmental accounts/edited by Valeria Costantini, Massimiliano Mazzanti and Anna Montini.
p. cm.
Includes bibliographical references and index.
1. Environmental auditing. 2. National income–Accounting. 3. Environmental economics. I. Costantini, Valeria. II. Mazzanti, Massimiliano. III. Montini, Anna.
TD194.7.H93 2011
363.73′6–dc23
2011023141

ISBN 978-0-415-59421-9 (hbk)
ISBN 978-1-138-80760-0 (pbk)
ISBN 978-0-203-15351-2 (ebk)

Typeset in Times New Roman
by Wearset Ltd, Boldon, Tyne and Wear

La science manipule les choses et renonce à les habiter. Elle s'en donne des modèles internes et, opérant sur ces indices ou variables les transformations permises par leur définition, ne se confronte que de loin en loin avec le monde actuel.

Maurice Merleau-Ponty (*L'œil et l'esprit*, 1964)

Contents

Figures

Tables

Contributors

Anastasia Basina studied Business Administration (2006) at the University of Macedonia, Thessaloniki, Greece. She holds a Master's degree in Economics (2009) from the Department of Public Administration of Panteion University, Athens. During her postgraduate studies, she specialized in the field of renewable energy technologies and their impact on the environment and the Greek economy. She is also a member of the Greek Chamber of Commerce and since March 2010 has been responsible for the department of economics in the field of PV-Systems, in one of the pioneering solar and wind system companies in Greece.

Elisa Bonazzi has a Master's Degree in Economics from the University of Bologna and an MSc in Environmental Economics from Bocconi University, Milan. She is a PhD student in Agriculture Economics at the University of Bologna and research fellow at the Faculty of Agriculture, funded by Arpa (Environment Agency of Emilia-Romagna Region). Her main research studies focus on environmental economics and accounting.

Renato Marra Campanale is a researcher at the Italian National Institute for Environmental Protection and Research (ISPRA). He worked for a time at the OECD and cooperates with ISTAT on environmental accounting. He has a Master's degree in Environmental Policy from Pavia University, Italy.

Valeria Costantini is Assistant Professor at the Department of Economics, Roma Tre University (Italy) and Lecturer in Environmental Economics and Climate Change, and Urban Economics. She has worked as an economist at the Italian National Research Institute for New Technologies, Energy and Environment (ENEA). She is involved in several national and international research projects, with specific interest in environmental economics and climate change, international energy markets and innovation in green technologies.

Silvana Dalmazzone (PhD in Environmental Economics and Environmental Management, University of York, UK, 1999) is Assistant Professor of Economic Policy at the University of Turin, Italy, where she teaches environmental and natural resource economics. She is member and former director of the Interdisciplinary Research Institute on Sustainability (IRIS), Turin. She has co-edited *Governing the Environment: Salient Institutional Issues* (2009), *Environmental Governance and Decentralization* (2007), *The Economics of Biological Invasions* (2000) and authored numerous articles in volumes and journals in the areas of economics of biodiversity, resilience of economic-ecological systems, environmental Kuznets curve, biological invasions and environmental accounting.

Pablo del Río is tenured researcher at the Consejo Superior de Investigaciones Científicas (CSIC). He received his PhD in Economics from the Universidad Autónoma de Madrid in 2002. His research focuses on the factors influencing environmental technology change in firms, climate change mitigation measures and renewable energy support schemes. He has collaborated in several EU and national projects and has published his work in international journals.

Charalambos Economidis has a BSc in Economics from the University of Heidelberg, Germany in 1979 and a PhD in Economics from Panteion University, Athens in 1991. Since 2003, he has lectured in Economics at the Department of Public Administration (Economics Section) at Panteion University, where he has held the post of Associate Professor in Economics since 2008. He has been lecturing on postgraduate courses since 2005. His research interests include input–output analysis and environmental and air pollution. Since 1994 he has been a member of the International Input–Output Association. In 2007 he published a book entitled *Introduction to the System and the Analysis of Input–Output*. He has co-authored numerous papers in refereed journals and conference proceedings.

Aldo Femia has an MSc in Economics and Econometrics and a PhD in Political Economy and has worked at the National Research Council, the Wuppertal Institute fuer Klima, Umwelt und Energie and the OECD. He is currently Senior Researcher at the Italian National Institute of Statistics (ISTAT). His research interests are in the fields of sustainability, environmental accounting, material flow analysis and input–output analysis.

Alessandra La Notte (MA in Environmental Management and Development, Australian National University, Canberra Australia, 1998; PhD in Environmental Economics, University of Trento, Italy, 2004) is a researcher at the European Commission-Joint Research Centre and Contract Professor of Environmental Accounting at the University of Turin, Italy. Her research topics range from the economic valuation of ecosystem services to environmental and ecosystem accounting.

Giovanni Marin has been a PhD student in Economics at IMT (Institutions, Markets, Technologies) Advanced Studies Lucca (Italy) since 2009. He graduated in Applied Economics at the University of Ferrara (Italy) and was a Visiting Scholar at the Department of Economics at the University of California, Berkeley, in spring 2011. His current research focuses on analysis based on environmental accounts and on environmental patenting of European firms.

Massimiliano Mazzanti is Associate Professor in Economics and lecturer in Environmental Economics at the University of Ferrara. He is also a research collaborator with the CERIS DSE CNR Institute in Milan. His main research fields deal with environmental policy, economics of innovation, economic performance and innovation, economic evaluation by stated preference techniques, waste management and policy, climate change and development.

Anna Montini (PhD in Political Economics) is Assistant Professor in Economics and Lecturer in Economics and Environmental Economics at the University of Bologna. She is also a Research Fellow at the National Research Council in Milan (CERIS-DSE). Her main research interests concern environmental economics and policy, waste management and environmental-economic performances at geographical/spatial level.

José Manuel Rueda-Cantuche is Contractual Agent at the Institute for Prospective and Technological Studies (IPTS) at the European Commission's Joint Research Centre in Seville. He has conducted extensive research (1997–2008) on the compilation of supply–use and input–output tables for Andalusia and the compilation of time series of supply–use and input–output tables for individual member states, the euro area and the European Union for Eurostat (2007–). He was the editor of the International Input–Output Association's Newsletter for three years (2007–2010) and is currently one of its Council Members (2007–); he is also Vice-President of the Hispanic-American Input–Output Society (2008–).

Michele Sansoni is a Management Engineer and MSc in Geographic Information Systems and a project manager at Arpa Emilia-Romagna. An expert in environmental assessment, energy–environment interactions and GIS analysis, his work also focuses on the development of RAMEA matrices and the analysis of methodologies for conducting GHG inventories at local level.

Milan Ščasný is a Senior Research Fellow at Charles University Environment Center, Prague. His research activities cover several areas in the field of environmental economics including, *inter alia*, the valuation of non-market goods, especially health risks, consumer behaviour, distributional effects of environmental regulation and economic impact modelling. He has been involved in more than 15 research projects funded as part of the FP5 to FP7 programmes of the European Commission, most of which he has coordinated, and he also took part in the OECD project on Household Behaviour and Environmental Policy.

Athanasios Sfetsos received a BSc in Physics from the University of Patras in 1995. He obtained a PhD in Electrical Engineering from Imperial College, University of London. He collaborated with the Research Tourism Institute and the Center for Renewable Energy Studies prior to joining the Departmento de Engenharia Electrotécnica at the Nova Universidade de Lisboa. He is currently a research scientist with the Environmental Research Laboratory at NCSR Demokritos. His research interests include environmental risk management and mitigation and air pollution. He has co-authored more than 100 papers in refereed journals and conference proceedings.

Peter J. Stauvermann is Professor of Economics at Changwon National University (Republic of Korea). He obtained his PhD from the Technical University of Dortmund (Germany) and has worked at several other universities including the University of Rostock (Germany), the Technical University of Twente (the Netherlands), the University of Tilburg (the Netherlands) and the University of the South Pacific (Fiji). His fields of research are environmental economics, growth theory, regional economics and the economic theory of conflicts.

Miguel Ángel Tarancón has a PhD in Economics and is Associate Professor of Economic Statistics at the University of Castilla-La Mancha. His field of specialization is input–output analysis. Within this field, he has contributed to key international journals including *Economic Systems Research*, *Energy Economics*, *Energy Policy*, *Energy* and *Transportation Research*. He is also the author of a book on input–output (in Spanish) and has collaborated in other international publications.

Fusako Tsuchimoto Menkyna has been a Research Fellow at Charles University Environment Center, Prague, since 2009. She obtained a bachelor's degree in Law at the University of Tokyo in 2004 and an MA in Economics from CERGE-EI. She expects to obtain a PhD in Economics from CERGE-EI in 2011. Her fields of interest are Applied Econometrics, Environmental Economics, Law and Economics, Political Economics. In the field of Environmental Economics, she conducts research on empirical analysis on the effect of firm behaviour on environmental quality.

Angelica Tudini (MSc in Environmental and Resource Economics and MA in Economics and Statistics), is a senior researcher at the Italian National Statistical Institute (ISTAT) where she has been working on Environmental Accounting since 1995. Her main area of expertise at national and international level is in Integrated Economic and Environmental Accounts and specifically NAMEAs and environmental taxes.

Giusy Vetrella, MA in Economics and Statistics and Researcher at the Italian National Statistical Institute (ISTAT), has worked on Environmental Accounting since 1999. Her main area of expertise is the calculation, at national and regional (NUTS 2) level, of air emission accounts (NAMEA) and energy use tables in physical units. She is also an expert in the calculation of general government expenditure for environmental protection.

Abbreviations

ACID	Acidification
AGE	Applied General Equilibrium
Arpa	Agenzia Regionale per la Protezione dell'Ambiente (Regional Environmental Protection Agency, Emilia-Romagna, Italy)
ASIA	Archivio Statistico delle Imprese Attive (Statistical Inventory of Active Firms)
ATECO	classificazione delle ATtività ECOnomiche (Statistical Classification of Economic Activities in Italy)
CBS	Central Bureau of Statistics of the Netherlands
CEPA	Classification of Environmental Protection Activities
CF	carbon footprint
CH_4	methane
CLRTAP	(UNECE) Convention on Long-range Transboundary Air Pollution
CO	carbon monoxide
CO_2	carbon dioxide
COICOP	Classification of Individual Consumption According to Purpose (UN)
CORINAIR	CORe INventory AIR emissions
DTA	domestic technology assumption
EA	Environmental Accounts
EC	European Community
EE-IO	environmentally extended input–output
EE-IOA	environmentally extended input output analysis
EE-MRIO	environmentally extended Multi-Region Input–Output
EIPRO	Environmental Impact of Products
EKC	environmental Kuznets curve
EMEP	European Monitoring and Evaluation Programme
ENEA	Italian National Agency for Energy
EPA	Environmental Protection Agency
EPO	European Patent Office
ESA	European System of National and Regional Accounts
ESEA	European Strategy for Environmental Accounting

EU	European Union
EU27	European Union (27 member states)
Eurostat	Statistical Office of the European Community
GDP	gross domestic product
GHG	greenhouse gas
GNI	green national income
GVA	gross value added
GWP	Global Warming Potential
ICLEI	International Council for Local Environmental Initiatives
IDA	Index Decomposition Analysis
IEA	International Energy Agency
INE	Instituto Nacional de Estadistica (National Statistical Institute, Spain)
INEMAR	INventario EMissioni Aria (Inventory of Air Emissions)
I–O	input–output
IPC	International Patent Classification
IPCC	Intergovernmental Panel on Climate Change
IPTS	Institute for Prospective Technological Studies
IREA	Inventario Regionale delle emissioni in atmosfera (Regional inventory of Air Emissions)
IRPET	Istituto Regionale per la Programmazione Economica della Toscana (Tuscan Regional Institute for Economic Planning)
ISTAT	Istituto di Statistica (National Statistical Institute, Italy)
IV	instrumental variables (estimator)
LM	Lagrange Multipliers (test)
LMDI	Logarithmic Mean Divisia Index
MRIO	Multi-Region Input–Output
MUD	Modello Unico di Dichiarazione ambientale (Declaration of Industrial Waste Production)
N_2O	nitrous oxide
NACE	Nomenclature statistique des activités économiques dans la Communauté européenne (Statistical Classification of Economic Activities in the European Community)
NAM	National Accounting Matrix
NAMEA	National Accounting Matrix including Environmental Accounts
NGO	non-governmental organization
NH_3	ammonia
NMVOC	non-methane volatile organic compounds
NNI	net national income
NO_x	nitrogen oxides
NSI	National Statistical Institute
NTNU	Norwegian University of Science and Technology
OECD	Organisation for Economic Cooperation and Development
OLS	Ordinary Least Squares
PAE	Potential Acid Equivalent

PATSTAT	Worldwide Patent Statistical Database
PM	particulate matter
PM_{10}	particulate matter (up to 10 micrometers in size)
PPC	Public Power Company
PSUT	physical supply-and-use table
RAMEA	Regional Accounting Matrix including Environmental Accounts
REGPAT	(OECD) Regional Patent Database
RES	Renewable Energy Sources
REZZO	Air Pollution Emission Source Register
RoW	Rest of the World
RUG	University of Groningen
SAM	satellite accounting matrix
SCP	sustainable consumption and production
SDA	structural decomposition analysis
SEEA	System of integrated Environmental and Economic Accounting
SEEA-E	System of Environmental-Economic Accounting for Energy
SMEs	small and medium enterprises
SNA	System of National Accounts
SNAP	Selected Nomenclature for Air Pollution
SNI	sustainable national income
SO_2	sulphur dioxide
SO_x	sulphur oxides
TERNA	Italian company responsible for electricity transmission
TNO	Netherlands Organization for Applied Scientific Research
toe	tons of oil equivalent
TOFP	Tropospheric Ozone Forming Potential
TSAP	Thematic Strategy on Air Pollution
UNECE	United Nations Economic Commission for Europa
UNFCCC CRF	Common Reporting Format
UNFCCC	United Nations Framework Convention on Climate Change
VIF	Variance Inflation Factor
WBCSD	World Business Council for Sustainable Development
WIPO	World Intellectual Property Organization
WRR	Scientific Council for Government Policy of the Netherlands
WWF	World Wide Fund for Nature
ZEP	Zero Emission Fossil Fuel Power Plants

Introduction

Valeria Costantini, Massimiliano Mazzanti and Anna Montini

This book is an integrated collection of complementary essays that revolve around the issue of environment-economic accounting, an analysis of which is crucial to assessing the static and dynamic performance of economic systems. Environmental and industrial policies also receive support from this framework of analysis, especially due to the detailed sector-based knowledge that settings such as NAMEA-type tables (National Accounting Matrix including Environmental Accounts) and Input–Output (I–O) provide. We aim to highlight studies focusing on the analysis of national or regionalized NAMEA through econometric and decomposition analyses and studies that try to link and merge NAMEA and I–O tables. To some extent, the book starts from and builds on what was explored in Part II of Mazzanti and Montini's *Environmental Efficiency, Innovation and Economic Performance* (2010).

The generation of richer and longer NAMEA datasets, both in the sectoral and temporal directions, has been matched by an increasing analytical effort over the last decade. Contributions by de Haan (2004), de Haan and Keuning (1996), Mazzanti and Montini (2010b) and Mazzanti *et al.* (2008) have emphasized the usefulness of NAMEA datasets for econometric investigations covering a number of different economic aspects.

It is worth noting that Eurostat has intensified its commitment to reach a full EU27 NAMEA, which has been released in 2011. It covers EU27 countries on 2000–2006 and is primarily aimed at sustaining EU research efforts and policy-making in the joint realms of resource efficiency (RE) and sustainable consumption and production (SCP), two EU strategic environmental pillars. This effort is considered a silver bullet of the EU strategy on data generation and policy support since it is recognized as a powerful instrument for assessing sustainable production and consumption performances (Watson and Moll, 2008; Moll *et al.*, 2007; EC, 2011; Eurostat, 2011). In line with the increasing EU emphasis on resource efficiency and decoupling targets, this 'economics of SCP' framework also includes as far as possible environmental innovation and its diffusion at sector and geographical levels as a key element of understanding (Kemp and Pearson, 2007; Popp, 2002). The research directions we deal with therefore offer a macro/meso perspective to sustainability.

This integrated EU data production adds to the historical national NAMEA dataset generation that in Italy, for example, is currently presenting a full

1990–2008 NAMEA with 60 NACE economic activities and three household categories. A new wave in data generation regards the 'regionalization' of the NAMEA which has brought new possibilities to the research ground since it considers both regional and sector dimensions. These regionalized NAMEAs maintain many different pollutants or aggregated environmental themes differentiated by their geographical distribution such as a more global climate change issue or a more localized acidification process. Interesting results may arise from application to a regional NAMEA as far as the role of innovation spillovers and environmental externalities on behaviours and location decisions by economic agents is concerned.

On the other hand, the integration of NAMEA-type tables and input–output (I–O) tables is a challenging but promising way to analyse the factors behind income–environment relationships in international settings (Cole, 2004; Copeland and Taylor, 2004; Frankel and Rose, 2005). More specifically, it allows disentangling of production and consumption perspectives, as far as the air emissions context is concerned,[1] through detailed sector-based information provided by the two frameworks. National and international sources of environmental effects can be ascertained in strict connection with streams of literature such as the ecological footprint kind of analysis and trade-oriented decomposition analyses.

Many types of analysis on the relationship between economic systems and the environment identify structural changes in production and consumption patterns as key drivers of environmental performance ('composition effect' in the environmental Kuznets curve (EKC) literature; see Mazzanti and Montini, 2010a). In addition to more qualitative assessments of the role of structural change in consumption and production, input–output analysis is a powerful analytical tool for investigating the role of change in the composition of final demand and the structure of intermediate inputs in determining aggregate environmental performance.

When specifically looking at international issues revolving around the environmental sustainability of countries, the integration between NAMEA and I–O is meaningful and even necessary when investigating to what extent changes in final consumption patterns, production technologies and trade patterns (as a result of the decoupling of consumption from production) affect domestic and world-induced air emissions. Furthermore, input–output analysis allows us to quantify to what extent geographical separation between consumption and production activities has occurred and whether it has determined increases (following the pollution haven hypothesis) or decreases in global environmental pressures. An analysis of the environmental pressures induced by vertically integrated sectors can be used to identify which categories of final demand were responsible for global environmental pressures.

This collection of essays tries to capture the aforementioned developments as far as possible, guiding the reader through the many specific as well as heterogeneous issues related to hybrid economic-environmental accounts.

The book is structured in two parts. After a historical background and empirical issues related to the NAMEA-type table definitions and estimations provided by Tudini and Vetrella and Stauvermann, the first part presents some applications and analyses mainly applied to the Italian context, with an insight into advancements made with regard to the regional NAMEA. The first chapter by Tudini and Vetrella (Chapter 1) sets out the methodological fundamentals of NAMEA. Its weaknesses offer room for improving analysis and a rationale for integration with other sources. The following chapter by Stauvermann (Chapter 2) presents the rationale behind the two possible approaches in the environmental accounting system (with some reference to the Dutch case), with or without monetization of the environmental damage, by comparing green accounting with a hybrid accounting approach.

Costantini *et al.* (Chapter 3) present the results of the first econometric-oriented analysis on the newly released Italian regional NAMEA. Their analysis of potential drivers for explaining the geographical distribution of environmental performances, greenhouse gases and acidification emissions reveals that technological and environmental efficiency spillovers are highly relevant. Bonazzi and Sansoni (Chapter 4) on the one hand, and Dalmazzone and La Notte (Chapter 5) on the other, offer further analyses by using regionalized NAMEA accounting for Emilia-Romagna, Piedmont and Lombardy, three major Northern Italian regions. These works extend the preliminary analysis of the 2000 Lazio regional NAMEA proposed by Mazzanti and Montini (2010b). The set of regional NAMEA works highlights the skyrocketing properties of such data frameworks for joint economic, environmental, regional and innovation studies.

A decomposition analysis applied to new Italian data performed by Femia and Marra Campanale follows in Chapter 6. They find that, in the Italian case, the amount of greenhouse gas (GHG) emissions displaced abroad, although not high in relative terms, denotes a significant displacement of the Italian GHG emissions towards the rest of the world, due to the growing share of imports in the final demand for products of the manufacturing industries and in the intermediate demand of these industries. If we care about the global climate, displaced emissions must also be measured and considered in global and national policies and targets.

The last work conceptually links Parts I and II of the book insofar as it deals with international and trade issues that are important when studying sustainable consumption and production dynamics.

Part II opens the window to international case studies for different EU countries and studies with methodological insights. We cover a few EU countries (Italy, Spain, Greece, Czech Republic).

Rueda-Cantuche (Chapter 7) develops a theoretical multi-regional framework (in two regions, Europe and the Rest of the World) with which the European carbon footprint can be calculated under the ideal conditions of full information availability. He shows three different approaches based on assumptions essentially related to production technology and emission coefficients by region. He concludes that the assumption of domestic emission coefficients has serious consequences for the estimation of the European carbon footprint mainly due to the fact that it does not capture the changes in the import shares of intermediate and/or final uses. The chapter by Marin *et al.* (Chapter 8) discusses how to join I–O and NAMEA from a conceptual point of view and the potentialities for research and policy-making. They also present new evidence on Italy and Spain with the aim of disentangling production and consumption environmental responsibilities, according to different levels of productive sectors aggregation, thereby addressing the so-called 'aggregation bias'. Their empirical findings, for the Italian and the Spanish cases, show that different sectoral aggregations produce significantly biased estimations of the amount of emissions both from a consumption perspective and a production perspective. This result suggests that particular attention should be paid to the interpretation of environmentally extended input–output analyses (EE-IOA) of country estimated amounts of embodied emissions, both in the domestic final demand and in those directly associated with the productive sectors, when the sectoral aggregation level has a low definition as considered in some recent similar studies. The work is the final version of a research effort that began with a presentation at the international conference The Structure of Economic Systems through Input–Output Applications (Accademia Nazionale dei Lincei, Rome, 21–22 October 2010).

Tarancón and del Río (Chapter 9) present a methodological study with a classification of applications of the input–output model applied to environmental issues in the context of changes in input–output coefficients. They analyse the variability of the coefficients and review different techniques to analyse the sensitivity of NAMEA variables to changes in the productive structure of activity branches.

Basina *et al.* (Chapter 10) analyse the emissions caused by conventional energy production methods and, specifically, by the combustion of lignite to generate electricity in Greece. They also analyse a reduction in emissions brought about by the possible replacement of lignite with solar energy in the production of electricity and, specifically, the use of photovoltaic technology. Calculations are made on the basis of the 2005 NAMEA table for the Greek economy as well as the domestic operational programme for the period 2007–2012 which considers the country's compliance with Kyoto Protocol requirements.

A lens on Eastern Europe is proposed by Ščasný and Tsuchimoto (Chapter 11), who utilize index-based statistical decomposition to examine which factors were active in changing the emission level of three pollutants, SO_2, NO_x and PM, during the transition and post-transition periods of the Czech economy (1997–2007). With a four-factor decomposition analysis on the year-by-year changes, they are able to distinguish the abatement effect due to output reduction, the one due to fuel intensity changes and lastly the effect of abatement reached through the installation of end-of-pipe technologies or through a fuel switch towards more environmentally friendly energy carriers.

Finally, we would like to thank all the contributors and colleagues who have taken part in this specific research effort and have worked intensively with us over the past years. Many of them are authors in this book.[2] Of those who have not participated in this book, we would particularly like to thank Roberto Zoboli. Roberto has always enriched our research with his brilliant and stimulating way of thinking and suggestions on how to interpret environmental and economic phenomena. This is the proof that, rather than a collection, this work is a tangible demonstration of the research efforts and ideas that we have tried to develop and exchange. We believe that this book is a platform that can guide us towards new research achievements and ideas in an even brighter future for environmental and ecological economics studies. Given that this book talks about hybrid phenomena and connections between the economic and environmental worlds, Massimiliano Mazzanti wishes to remember Alexander Langer, a man who devoted his life to creating links and bridges between apparently different realities, languages and cultures in order to demonstrate the richness and value of hybrid and mixed things in the human world.

Notes

1 Future developments in NAMEA-type tables involve water and soil emissions. A specific issue of 'dispersion of toxic substances' that also covers pollution of water and soil by pesticides and other substances, such as heavy metals and dioxins, addresses the methodological problems to be solved.

2 It is worth noting that some of the authors presented papers that dealt with NAMEA/I–O use at the last I–O Society Conference in Sydney 2010. José Rueda-Cantuche won the Leontief prize there.

References

Cole, M.A. (2004) 'Trade, the pollution haven hypothesis and the environmental Kuznets curve: examining the linkages', *Ecological Economics*, 48(1): 71–81.

Copeland, B.R. and Taylor, M.S. (2004) 'Trade, growth and the environment', *Journal of Economic Literature*, 42: 7–71.

de Haan, M. (2004) 'Accounting for goods and for bads: measuring environmental pressure in a national accounts framework', PhD thesis, Universiteit Twente.

de Haan, M. and Keuning, S.J. (1996) 'Taking the environment into account: the NAMEA approach', *Review of Income and Wealth*, 42(2): 131–148.

European Commission (EC) (2011), *Roadmap to a Resource Efficient Europe*, Bruxelles.

Eurostat (2011), Creating consolidated and aggregated EU27 Supply, Use and Input-Output Tables, adding environmental extensions (air emissions), and conducting Leontief-type modelling to approximate carbon and other 'footprints' of EU27 consumption for 2000 to 2006, EUROSTAT, Luxembourg.

Frankel, J. and Rose, A.K. (2005) 'Is trade good or bad for the environment? Sorting out the causality', *Review of Economics and Statistics*, 87: 85–91.

Kemp, R. and Pearson, P. (2007) *Final Report of the MEI Project Measuring Eco Innovation*, Maastricht: UM MERIT.

Mazzanti, M. and Montini, A. (2010a) *Environmental Efficiency, Innovation and Economic Performance*, Abingdon: Routledge.

Mazzanti, M. and Montini, A. (2010b) 'Embedding emission efficiency at regional level: analyses using NAMEA', *Ecological Economics*, 69(12): 2457–2467.

Mazzanti, M., Montini, A. and Zoboli, R. (2008) 'Environmental Kuznets curves for air pollutant emissions in Italy: evidence from environmental accounts (NAMEA) panel data', *Economic Systems Research*, 20: 277–301.

Moll, S., Vrgoc, M., Watson, D., Femia, A., Pedersen, O.G. and Villanueva, A. (2007) 'Environmental input–output analyses based on NAMEA data: a comparative European study on environmental pressures arising from consumption and production patterns', ETC/RWM working paper 2/2007, Copenhagen: European Topic Centre – Resource and Waste Management.

Popp, D. (2002) 'Induced innovation and energy prices', *American Economic Review*, 92: 160–180.

Watson, D. and Moll, S. (2008) 'Environmental benefits and disadvantages of economic specialisation within global markets, and implications for SCP monitoring', paper presented at the SCORE! conference, 10–11 March 2008, Brussels, Belgium.

Part I

The foundations of NAMEA and recent developments

Methods and analysis

1 NAMEA

From pioneer work to regulation and beyond

Angelica Tudini and Giusy Vetrella

Introduction

NAMEA (National Accounting Matrix including Environmental Accounts) is a statistical framework which extends the matrix presentation of national accounts (the NAM) to environmental flows. The term originates from the work conducted by the Dutch Statistical Institute in the 1990s.[1] Early versions of the Dutch NAMEA were very comprehensive on the NAM side and covered income generation, distribution and use accounts as well as accumulation accounts and changes in balance-sheet accounts; environmental flows mainly included the generation of air pollutants, solid waste, phosphorus and nitrogen by production activities and household consumption. Over time and with the United Nations handbook *System of Integrated Environmental and Economic Accounting* (SEEA[2]), in particular, the term NAMEA became used as a synonym for hybrid flow accounts i.e. a matrix framework which presents an economic module including national economic accounts in monetary terms, (the NAM) side by side with an environmental module including flow accounts in physical units (EA), both modules being based on common national accounts principles; the term hybrid refers to the joint use of monetary and physical units.

The second section of the chapter describes a Hybrid Flow Account where the economic module is in the form of monetary Supply and Use. On the environmental side, the development of NAMEA-type accounts started in most EU countries from air emissions, thanks to the availability of good and comparable statistical data sources. Now referred to as air emission accounts, they are one of the three modules for which the forthcoming EU Regulation on environmental accounts will make it compulsory to produce nation-wide yearly time series. The other two modules are economy-wide material flow accounts and environmentally related taxes by economic activity.

In Italy, the Italian National Statistical Institute (ISTAT) has been regularly releasing a time series of air emission accounts at national level since 2004; air emission accounts at regional (NUTS 2) level for 2005 are also available.

The emphasis given to the sub-national breakdown of data is more the result of national priorities than the directions agreed upon at EU level, where the foreseen short-term development for NAMEA-type accounts is not in the direction of a spatial or temporal breakdown, but aims to expand the set of environmental pressures, particularly with energy accounts (in the medium term), and possibly water accounts (in the longer run).

The compilation of NAMEA-type accounts on the basis of existing statistics, which were not traditionally designed to be consistent with monetary economic accounts, requires a number of steps, described in the third section of the chapter, mainly with reference to air emission accounts and energy accounts.

NAMEA accounts and tables

A widely known kind of hybrid flow account[3] is based on the integration of a monetary supply-and-use table for the economic module and physical flow accounts for the environmental module. The economic and environmental components are described below, first as two separate entities (Figure 1.1 and Figure 1.2) and then merged in one single matrix framework (Figure 1.3).

Figure 1.1 presents a monetary supply-and-use table, a standard framework in national accounts. It shows the monetary value of products – goods and services – made available in the economy by means of domestic production or imports by producing industry and the use of products; it also shows the intermediate costs of economic activities as well as their value added (output minus intermediate consumption). The first row/column pair presents the goods and services account: column-wise (supply – resources) the supply of goods and services is shown first, with a distinction made between domestic industry products[4] and imports; the column also records trade and transport margins and net indirect taxes charged on products thus ensuring that supply and use totals match.[5] Row-wise, the account shows all possible uses of available resources: intermediate consumption by industries, final consumption, gross capital formation and exports.

Figure 1.2 presents a pair of physical supply-and-use tables (PSUTs); they describe the origin and the destination of three kinds of flows:[6] (a) resources –

	Products	Industries	Final uses			Total
Products		Intermediate consumption	Final consumption	Gross capital formation	Exports	Total use by product
Industries	Output by product and by industry					
Imports	Products imported					
Margins	*Trade and transport margins*					
Net taxes on products	*Taxes less subsidies on products*					
Value added		*Value added by industry*				
Total	Total supply (output + imports) by product					

Figure 1.1 Monetary supply-and-use table.

minerals, energy resources, water and biological resources; (b) products – goods and services produced within the economic system and also represented in Figure 1.1; (c) residuals – solid, liquid and gaseous residuals.

The physical supply table shows the origin of all flows (row headings) that can occur between the economy and the environment: by definition, natural resources can only be supplied by the environment whereas products and residuals can only be supplied by the economy, either the domestic industry or the rest of the world (RoW) in the form of imports.[7] The product flows portrayed in the table are the physical counterpart of monetary transactions in Figure 1.1. In the full tables, each flow category is broken down according to a suitable existing classification, for example, for energy resources, the breakdown follows the classification of energy resource assets proposed in the draft System of Environmental-Economic Accounting for Energy (SEEA–E[8]). Column headings may also be detailed, particularly for industries, broken down according to the standard industry classification, NACE (Statistical Classification of Economic Activities in the European Community, version Rev. 1.1).

The use table shows, row-wise, the uses of available flows by industry for intermediate consumption, final users including RoW (for exports) and the environment.

Supply (origin)

		Economy		Environment
		Industries	**RoW**	
Flows	(i) resources			
	(ii) products			
	(iii) residuals			

not applicable

Use (destination)

		Economy			Environment
		Intermediate consumption (by industry)	**Final consumption**	**RoW**	
Flows	(i) resources				
	(ii) products				
	(iii) residuals				

not applicable

Figure 1.2 Physical supply-and-use tables.

All flows are quantified in physical natural units (tonnes, cubic metres, etc.). As in monetary tables, total supply also equals total use (for each flow) in physical tables.

The rationale of hybrid accounts is to represent the economy–environment interface by applying the framework of supply-and-use tables (which scores high in terms of analytical potential) to flows other than product flows while maintaining different units (monetary and physical) in the scheme.

Figure 1.3 is a hybrid flow account obtained by merging the monetary supply-and-use tables in Figure 1.1 (economic module) with the PSUTs in Figure 1.2 (environmental module).

The economic module presents one more element than Figure 1.1: household consumption according to heating, transport and a residual 'other' category; these items, representing an 'of which' of final consumption, are particularly meaningful in the case of air emission accounts and energy accounts since they show the specific household consumption items associated with air emissions and energy use.

The environmental module describes the origin (columns) and destination (rows) of 'natural resources' and 'residuals' in one single framework as in Figure 1.2; by contrast, physical flows of products are not included in Figure 1.3 since the description of the flows of goods and services within the economy is provided in monetary terms in the economic module.

For natural resources, the accounting framework describes their supply by the environment and their use by production activities (industries), consumption activities and exports.

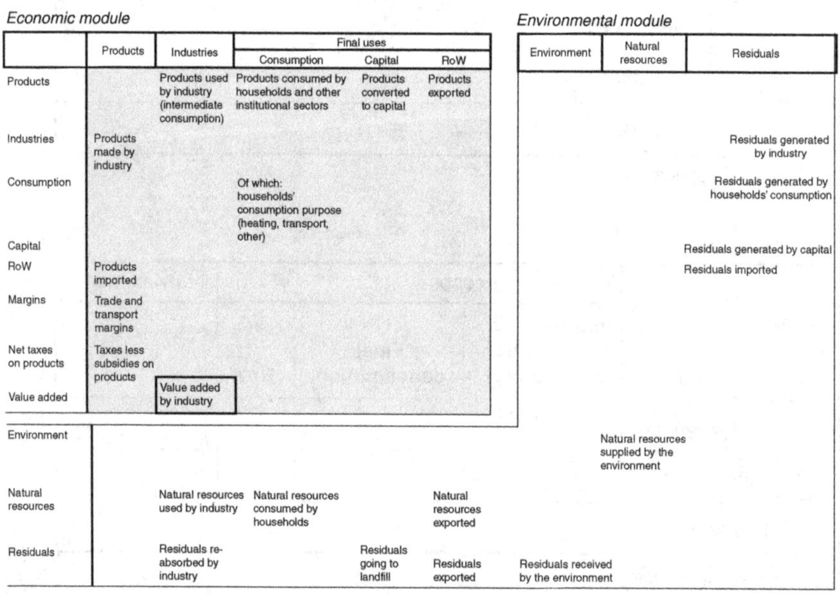

Figure 1.3 Supply-and-use-based NAMEA (source: United Nations *et al.*, 2003, § 4.38).

As regards residuals, the accounting framework describes their origin/supply – production activities (industries), consumption activities (private households), capital and imports – as well as their destination/use, which includes their use as input for industries, as capital stock (residuals going to landfill) and as exports. The balance between residuals supplied by the economy and those used by the economic system itself is the quantity of residuals released to the environmental system.[9]

The joint presentation of Figure 1.3 is typical of the NAMEA framework with the environmental module reporting the environmental pressures generated by the economy (air emissions, use of natural resources, etc.) and the economic module accounting for the socio-economic parameters (production, value added, employment, etc.) corresponding to the economic activities (industries and households) that generate environmental pressures. With regard to industries, the framework identifies two joint results generated by the activity carried out for each economic sector: economic values (output, value added, etc.) and pressures exerted to generate the economic values themselves. As far as households are concerned, the pressures generated by selected consumption activities are compared with the expenses incurred by households to purchase products whose use is at the root of the environmental pressures themselves.

For both production activities and households, environmental pressures are allocated to who is directly responsible for their generation (due to production or consumption processes respectively for industries and households).[10] Alternatively, NAMEA data can be further processed using input–output analysis to give a consumption perspective, in which air emissions are reattributed to the production chains of final products.[11]

The acronym NAMEA is also used for data tables rather than matrix frameworks. Figure 1.4 presents a typical NAMEA-type table: columns cover the main

Economic activities and household consumption	Economic data			Environmental pressures	
	Value added	Employees	Household's consumption	Air emissions	Energy use
Agriculture			▒		
...			▒		
Manufacturing industry			▒		
Energy			▒		
...			▒		
			▒		
...			▒		
			▒		
...			▒		
Household consumption – Transport – Heating – Other	▒	▒			

▒ Not applicable

Figure 1.4 NAMEA tables.

socio-economic aggregates as well as environmental pressure data (e.g. air emissions, energy use, water use, waste generation) broken down (rows) into industry (NACE classification) and household consumption function.

NAMEA-type tables according to the Figure 1.4 format are regularly available in Italy for air emissions.[12] In addition to air emission accounts at national level, regional (NUTS 2) level air emission accounts are available for 2005.[13]

With regard to energy accounts, the release of ISTAT's energy use data by economic activity for the years 1990–2008 is forthcoming whereas the production of complete physical supply-and-use tables for energy will be coordinated with Eurostat's schedule.

At EU level, air emission account data are collected by Eurostat via bi-annual questionnaires – covering the years 1995 to t–2, 13 air pollutants[14] and 60 NACE (in its version Rev 1.1) industries plus three household consumption categories – addressed to all EU member states and Norway and Switzerland.[15] Since data collection is not yet compulsory, responses by member states show a high variability in data coverage (among countries and pollutants).[16] The entry into force of the EU Regulation on environmental accounts will make it compulsory for member states to deliver data for 14 air pollutants[17] broken down into economic activities (A64 level of the NACE Rev. 2, the most recent version of NACE). Further developments for NAMEA-type accounts in the EU will be in line with the priorities defined for physical accounts by the revised European Strategy for Environmental Accounting (ESEA 2008): energy accounts for the medium term and water accounts in the longer run.[18] In both cases, data production will start from physical flow accounts by adopting the framework of physical supply-and-use tables described in Figure 1.2.

The explanation of the main features of the NAMEA framework, in this section, makes it clear that the main requirement of a NAMEA-type account is that environmental data are consistent with the national accounts principles and classifications that hold for the economic data.

Since environmental statistics used as primary data for environmental accounts are not compiled according to national accounts principles in the first place, environmental statistics need to be adjusted in order to be included in the NAMEA framework.

The next section describes the main issues that compilers face in practice when producing NAMEA-type accounts.

Compiling NAMEA-type accounts

This section explains the most common adjustments operated on environmental statistics when used in NAMEA-type accounts. Reference is made to the specific case of air emission accounts as well as energy accounts which are likely to pose similar problems. The kind of adjustment described here is common to all countries which derive air emission accounts from national emission inventories, i.e. national databases for air emissions generally used for countries' communications within the framework of international conventions: the United Nations Framework

Convention on Climate Change (UNFCCC) and the UNECE Convention on Long-range Transboundary Air Pollution (CLRTAP);[19] a very similar situation occurs when energy accounts are derived from energy statistics/balances.

Following the international guidelines on air emissions, national inventories cover emissions that broadly correspond to the (national) geographic territory; the geographic territory is also the reference for energy balances.

By contrast, consistency with national accounts requires that all flows relate to the activities of 'residents units' (this is known as 'residence principle'):

> an institutional unit is said to be resident within the economic territory of a country when it maintains a centre of economic interest in that territory – that is, when it engages, or intends to engage, in economic activities or transactions on a significant scale either indefinitely or over a long period of time, usually interpreted as one year. Some of the production of a resident institutional unit may take place abroad, while some of the production taking place within a country may be attributable to foreign institutional units.
>
> (Eurostat, 1996, § 1.30)

The adoption of the residence principle implies that air emissions/energy use in air emission accounts/energy accounts has to be related to resident units; its implementation means converting data from the geographic definition of a country's territory which is adopted in air emission inventories/energy statistics and balances to the economic definition which forms the basis for national accounts. Figure 1.5 illustrates that the conversion implies including the activities of

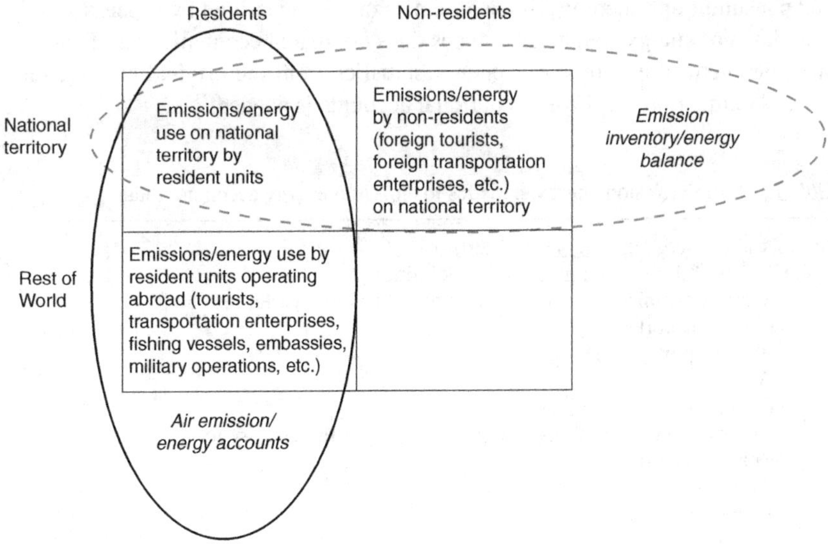

Figure 1.5 The coverage of resident and non-resident units in emission/energy statistics and accounts (source: adapted from Eurostat, 2009).

resident units in the Rest of the World (which are part of the economic definition but not included in the geographic one) and excluding the activities of non-resident units on the national territory (which are part of the geographic notion and do not fit into the economic one).

In practice, the adjustment is needed for economic activities that engage in international transport – by road, water and air – as well as for tourism consumption; for stationary economic activities (all except transport), the geographic and economic notions coincide.

The adjustment process needed to make the system boundary of environmental accounts consistent with the national accounts' one implies that totals derived from air emission accounts and energy accounts do not equal, respectively, the emission totals of the main international agreements on emissions of air pollutants[20] and the energy statistics totals.

Table 1.1 illustrates in detail how the emission/energy statistics totals and the accounts totals are related.[21]

Statistics needed to make the required corrections include, for example, fuel purchases abroad by tourists as well as by lorries, aircraft, ships belonging to resident companies, and the same purchases by non-residents on the country's territory. As fuel-purchase data in many countries are likely not to include all the details needed, other statistical sources such as transportation statistics or monetary figures can be used to estimate the necessary distinction between resident units and non-resident units.[22]

In addition to applying the residence principle, another important step in the implementation of air emission/energy accounts relates to the breakdown of flows into economic activity (i.e. according to NACE classification and household consumption function). Since air emissions are classified by process and the breakdown of energy statistics/balances does not match economic classifications, a reassignment of air emissions/energy statistics from the original classification to the classification employed in national accounts is needed.[23]

Table 1.1 From emission/energy statistics to emission/energy accounts' totals

Air emissions/energy use on national territory
(UNFCCC or CLRTAP/energy statistics definition)
- (+) Residents' emissions/energy use in the rest of the world, for:
  - Road transport
  - Air transport
  - Water transport
  - Fishing
- (−) Non-residents' emissions/energy use on domestic territory for:
  - Road transport
  - Air transport
  - Water transport
  - Fishing

= Air emissions/energy use by resident units (i.e. the National Accounts definition)

Source: adapted from Hass and Kolshus (2007).

Conclusions

From an EU-wide standpoint, the availability of NAMEA-type data is likely to improve as a result of the priorities set by Eurostat which are consistent with the recommendations provided by the 2008 European Strategy for Environmental Accounting (ESEA).

In the short term, the entry into force of the EU Regulation on environmental accounts will ensure the supply of a comparable set of air emission accounts time series up to the year t–2 for all member states.

In the medium term, the foreseen implementation of energy physical supply-and-use tables will bring advantages stemming not only from the high analytical potential of energy PSUTs per se, but also from the possibility of combining the new sets of energy flow accounts data with air emission accounts data.

Furthermore, the possible extension of the framework to water PSUTs in the longer run would make socio-economic, energy use, water use and air and water emissions data available with the same breakdown by industry (and household consumption function), thus significantly enhancing the scope of analysis and providing valuable input for environmentally extended macro-economic models.

Although it looks very promising, the coordinated development of NAMEA-type accounts foreseen at EU level does not fulfil all of the needs for improved data and indicators expressed by users in general and specifically by policy-makers.

Particularly relevant in this context are the expected innovations in statistical data at large which derive from projects that aim to measure societies' progress, namely the European Commission's proposed actions to improve the measurement of progress in a changing world, the ongoing OECD Global Project on Measuring the Progress of Societies and the recommendations of the 'Stiglitz–Sen–Fitoussi Commission' on the Measurement of Economic Performance and Social Progress. For environmental accounts in particular, the needs expressed by these projects include: (a) increasing the timeliness of environmental accounts, thereby reducing the current gap with national accounts' data; (b) addressing distributional issues; (c) providing data with a territorial and temporal breakdown which goes beyond the current standard: national for the territory and year as temporal reference.

Although no EU-wide efforts to respond to these policy needs in a coordinated manner have been recorded yet, significant examples exist of individual countries that fulfil these objectives in their national release of NAMEA-type accounts and derived indicators.

With regard to timeliness – as a highly sensitive policy issue since it would allow the provision of short-term integrated environmental and socio-economic information to policy-makers – countries like Germany and the Netherlands release NAMEA-type estimates for the year t_{-1}. The Netherlands is also in the forefront as regards distributional issues with its household air emissions account by income size and as regards the availability of data with a temporal breakdown, with its quarterly CO_2 emission accounts.

Italy is leading the way in the field of territorial breakdown with the production in 2009 of the first set of air emission accounts at regional level and is to our knowledge the only EU country engaged in this exercise so far. The priority given at ISTAT to the spatial breakdown is the answer to the demand for subnational data in a country with a high level of regional diversity in natural resource endowment as well as socio-economic and technological development.[24] The calculation of regional air emission accounts for additional years (planned for 2011)[25] is a necessary step which will allow more comprehensive analysis to be pursued also at regional level.

Notes

1 See de Haan and Keuning (1994).

2 SEEA is the main international reference for the analysis of the relationship between the environment and the economy in a satellite account form. Reference is made here to the 2003 version, known as SEEA2003, see for example United Nations *et al.* (2003). The SEEA is currently being revised and is expected to be raised by 2012 to the formal status of an international statistical standard.

3 See UN *et al.* (2003), SEEA2003, mainly figure 1.1 in § 4.38.

4 Measured at basic prices. The basic price is the price receivable by producers from the purchaser for a unit of a good or service produced as output minus any tax payable on that unit as a consequence of its production or sale (i.e. taxes on products), plus any subsidy receivable on that unit as a consequence of its production or sale (i.e. subsidies on products).

5 Measured at purchasers' prices. The purchaser's price is the amount actually paid by the purchaser per unit of goods or service bought. It includes taxes, with VAT counting only for its non-deductible part, and the subsidies on the products are deducted. It also includes the costs of transport paid separately by the purchaser to take possession of the products at the place and time required. It does not include any added interest if a loan is granted.

6 In addition to the three categories shown here, the SEEA2003 also includes a fourth category for 'ecosystem inputs': water and other natural inputs (e.g. nutrients, carbon dioxide) required by plants and animals for growth, and the oxygen necessary for combustion.

7 In practice, residuals can also be generated by the environment of the RoW in the form of transboundary flows (residuals transported by wind and water); since no explicit distinction is made here between the domestic and the RoW environment, the simplifying assumption holds that residuals can only be generated in the economic sphere.

8 The SEEA–E manual, which is currently being drafted, applies the more general environmental accounting principles and accounting frameworks outlined in the UN handbook System of Integrated Environmental and Economic Accounting (SEEA) to the specific domain of energy. See http://unstats.un.org/unsd/envaccounting/seeae.

9 In the specific case of air emissions, the amount of residuals used by the economic system (destination) is negligible; hence, total supply equals the amount of residuals released to the environmental sphere. This explains why air emission accounts tables only include items related to supply (see Figure 1.4).

10 This is the so-called 'production perspective' under which air emissions (e.g. due to electricity production) are entirely assigned to the producing industries and not to users.

11 See, for example, Marra Campanale and Femia's Chapter 6 in this book.

12 See ww.istat.it/conti/ambientali.

13 See ISTAT (2009). Data at the regional level were calculated by using as main input the emission inventory produced by the Istituto Superiore per la Ricerca e la Protezione Ambientale (ISPRA) at the NUTS 3 level, available for the years 1990, 1995, 2000 and 2005.

14 Carbon dioxide (CO_2), carbon dioxide from biomass, sulphur oxides (SO_x), nitrogen oxides (NO_x), nitrous oxide (N_2O), ammonia (NH_3), methane (CH_4), carbon monoxide (CO), non-methane volatile organic compounds (NMVOC), particulate matter (PM_{10}), perfluorocarbons (PFCs), hydrofluorocarbons (HFCs), sulphur hexafluoride (SF_6).

15 Results are published in the Eurostat database (http://epp.eurostat.ec.europa.eu/portal/page/portal/statistics/search_database). See also Eurostat (2010b). EU country groups (EU15, EU25, EU27) data are Eurostat estimates.

16 In Italy, for example, data are not reported for carbon dioxide from biomass, PFCs, HFCs, sulphur hexafluoride; only 9 out of the 19 pollutants covered by the annual air emission accounts time series can be used to fill in the Eurostat questionnaires (the other 10 are heavy metals which are not collected by Eurostat).

17 The 13 included in the voluntary questionnaires (see note 14) plus fine particles ($PM_{2,5}$).

18 For monetary environmental accounts, ESEA 2008 gives priority to resource use and management accounts as well as statistics on environmental goods and services. See Eurostat (2008).

19 Inventory-based air emission accounts are the most common ones at present; for detailed compilation suggestions applying to this case and also to countries that use different data inputs see Eurostat (2009). In Italy, the national emission inventory is produced by the ISPRA. Updated time series are available at www.sinanet.apat.it/it/sinanet/sstoriche.

20 The United Nations Framework Convention on Climate Change (UNFCCC) and the UN–ECE Convention on Long-Range Transboundary Air Pollution (CLRTAP) with reporting to UN–ECE/EMEP.

21 Table 1.1 compares UNFCCC/CLRTAP totals and air emission accounts totals. If the comparison is made between totals directly derived from the emission inventories and air emission accounts, in addition to the items listed in Table 1.1 a further possible element explaining the difference would be any emission from non-economic agents (e.g. nature) as well as nature's absorption of substances, covered in the national emission inventories and excluded in the accounts which only cover socio-economic related flows.

22 Compilation issues are widely analysed in the Eurostat Manual for air emission accounts (Eurostat, 2009). They are in most cases also relevant for energy accounts.

23 For details on how to assign process-based emissions to economic activities and households' consumption functions, see Eurostat (2009).

24 For details on the possible uses of regional NAMEA-type data in the specific context of development policies, see Cervigni *et al.* (2005).

25 The ISPRA emission inventory at the NUTS 3 level is currently available for 1990, 1995, 2000 and 2005; see: www.sinanet.apat.it/it/inventaria/disaggregazione_prov2005.

References

Cervigni, R., Costantino, C., Falcitelli, F., Femia, A., Pennisi, A. and Tudini, A. (2005) 'Development policies and the environment: using environmental accounts for better decision making', Materiali UVAL N. 5, Ministero dell'Economia e delle Finanze, Dipartimento per le Politiche di Sviluppo, Unità di Valutazione degli Investimenti Pubblici, Roma. Online: www.dps.tesoro.it/documentazione/uval/materiali_uval/MUVAL5_eng.pdf.

Commission of the European Communities (2009) *GDP and Beyond: Measuring Progress in a Changing World*, communication from the Commission to the Council and the European Parliament, COM(2009) 433 final, Brussels: European Commission.

de Haan, M. and Keuning, S. (1994) 'A national accounting matrix including environmental accounts: concepts and first results', paper presented at the UNEP workshop on Environmental and Resource Accounting, Slovakia, March 1994.

Eurostat (1996) *European System of Accounts: ESA 1995*, Brussels: European Commission.

Eurostat (2008) *Revised European Strategy for Environmental Accounting*, CPS doc. 2008/68/7/EN, Brussels: European Commission.

Eurostat (2009) *Manual for Air Emissions Accounts*, Brussels: European Commission.

Eurostat (2010a) 'Energy PSUTs: result tables (overview)', meeting of the Task Force 'NAMEA Energy Accounts', Brussels, 6–7 May 2010.

Eurostat (2010b) *Environmental Statistics and Accounts in Europe*, Brussels: European Commission.

Hass, J.L. and Kolshus, K.E. (2007) 'Requirements for energy statistics in linked environmental-economic data sets', second meeting of the Oslo Group on Energy Statistics, New Delhi, India, 5–7 February 2007.

ISTAT (2009) *Namea: emissioni atmosferiche regionali*, data table, Rome: ISTAT. Online: www.istat.it/dati/dataset/20090401_00.

Statistics Netherlands (2010) *Environmental Accounts of the Netherlands 2009*, Voorburg: Statistics Netherlands.

Stiglitz, J.E., Sen, A. and Fitoussi, J. (2009) *Report by the Commission on the Measurement of Economic Performance and Social Progress*. Online: www.stiglitz-sen-fitoussi.fr.

United Nations, European Commission, International Monetary Fund, Organisation for Economic Cooperation and Development and the World Bank (1993) *Handbook of National Accounting: Integrated Environmental and Economic Accounting*, Series F, No. 61, New York: Statistical Office of the United Nations.

United Nations, European Commission, International Monetary Fund, Organisation for Economic Cooperation and Development and the World Bank (2003) *Handbook of National Accounting: Integrated Environmental and Economic Accounting 2003* [SEEA 2003]. Online: http://unstats.un.org/unsd/envaccounting/seea.asp.

2 The NAMEA

Methodological questions and historical disputes in the Dutch experience

Peter J. Stauvermann

Introduction

The National Accounting Matrix including Environmental Accounts (NAMEA) is, from an economic history point of view, a relatively new approach which serves as an environmental-economic information tool for policy-makers, scientists and the interested public. It was mostly developed in the late 1980s and 1990s at the Central Bureau of Statistics of the Netherlands (CBS) because the Dutch public and policy-makers were aware of an increasing number of problems regarding the natural environment and wanted to get a deeper insight into the state of these environmental problems. Consequently, this work was delegated to the CBS, a public institution which is policy-independent and has a very good reputation. Within the CBS, there were two strands of methodology concerning what an information system of this type should look like; on the one hand, the advocates of a 'greened national income' (GNI) approach, represented by Roefie Hueting's department, and on the other hand, the advocates of a hybrid accounting system, represented by Steven Keuning's department. The main methodological difference was 'Does it make sense to monetize environmental damage and the environment?' This discussion was not only restricted to the CBS, it was extended to a discussion disputed by the scientific community in the Netherlands and in the Dutch parliament, and then at European and at worldwide level. It finally resulted in the SEEA (System of Economic and Environmental Accounts) 2003. We think this question is still an important one from an environmental-economic point of view and from a more general view related to how methodological disputes are solved. In this respect, it makes sense to go back 25 years, especially if we take into account, for example, discussion on the Stern report.

Here we want to give an overview of the advantages and disadvantages of both methodologies. In the next sections, we explain both approaches in a very simplified manner. We will also examine some critical points.

History

Since the environment began to play an increasingly important part in the focus of policy-makers and the public, the demand for environmental accounting increased in the 1980s. The motivation for environmental accounts was the adoption by governments of the notion of sustainable development together with the understanding that economic activities and appropriate economic incentives play a central role in determining whether development is environmentally sustainable or not. Environmental accounts shall provide policy-makers and the public, first of all, with indicators and descriptive statistics to monitor the interaction between the environment and the economy and, second, with a database for strategic planning and policy analysis in order to identify more sustainable development paths and policy instruments for achieving these paths.

In principle, there are two different methodologies used for accounting for the environment. On the one hand, some economists (Hueting, 1974, 1980; Mäler, 1991; Hartwick, 1990) propose adjusting the net national income (NNI) for the value of environmental damage to generate a GNI. On the other hand, other economists, mostly national accountants (Keuning, 1991; de Haan, 2004; Keuning and Steenge, 1999) only want to relate the economic performance of an economy to environmental damage measured in physical units. The main difference between both approaches is that supporters of the GNI propose monetarizing environmental damage whereas opponents of this approach only want to develop a hybrid accounting system. This scientific discussion took place at international level (London Group, OECD, World Bank, EU, statistical bureaus) and at the same time within the Netherlands. That means that the involvement of the Netherlands in the problem could be interpreted as a kind of mirror image which represents not only the development of environmental accounts within the Netherlands, but also development in the rest of the world.

Of course, the number of different definitions of a GNI seems to be endless, but in principle all these different concepts try to subtract environmental damage measured in monetary units from the conventional NNI. Only the methods used to value environmental damage are different.[1] In a certain sense, the approach of the Dutch economist Roefie Hueting (Hueting, 1974, 1980) seems to be representative of all other similar GNI approaches. To the author's knowledge, this was the first approach to be developed. Some other Dutch economists developed the idea of a hybrid accounting system which they called National Accounting Matrix including Environmental Accounts and is a so-called satellite system. The leading researcher was Steven Keuning. He and his colleagues developed an accounting matrix where they relate economic indicators (measured in monetary units) to environmental indicators (measured in physical units). Of course, there are also some alternative approaches such as material flow accounting, but the NAMEA seems to be superior to alternative approaches.[2] Because of the large number of existing books and papers in the literature on sustainability, we cannot review every approach. But the question that remains is, what does the term *sustainability* mean?[3]

This vague concept is related to a basic definition of sustainability in economics. The starting point of sustainability was the notion of sustainable income expressed by Hicks (1946): 'income is the maximum amount an individual can consume during a period and remain as well off at the end of the period as at the beginning.' The sustainable income of Hicks (1946) is interpreted as the amount of income that can be spent without depleting the wealth which generates income. Hence, sustainability requires non-decreasing levels of capital stock over time, or, at the level of the individual, non-decreasing per capita capital stock.

The indicators of sustainability could be based on either the value of total assets in every period or the change in wealth, depreciation of capital in the conventional national accounts. Consequently, for a proper measure of sustainability, all assets should be included in this type of indicator; manufactured capital, human capital and environmental capital.

In the last 50 years, only manufactured capital was recorded in the system of national accounts (SNA) because an accepted method to measure and value natural and human capital did not exist.[4]

Economic sustainability can be defined as strong or weak. The concepts of weak sustainability only require the aggregated value of all assets to remain constant. That means that one form of capital can be substituted for another and natural capital can therefore be depleted or the environment degraded as long as there are compensating investments in other types of capital: manufactured, human or other types of natural capital. In the words of Brekke (1997): 'A development is said to be weakly sustainable if the development in non-diminishing from generation to generation. This is by now the dominant interpretation of sustainability.' Common and Perrings (1992) called the concept of weak sustainability 'Hartwick–Solow sustainability'.[5]

In contrast to weak sustainability, the concept of strong sustainability requires each value of a specific form of natural capital to remain constant. The idea behind this strong concept is that natural capital is a complement to manufactured capital rather than a substitute. This concept of strong sustainability has some direct consequences for environmental policy: (a) renewable resources such as fish or forests can only be exploited at the natural rate of net growth; (b) the use of non-renewable resources should be minimized and, ideally, (c) used only at the rate for which renewable substitutes are available; (d) emissions should not exceed the assimilative capacity of the environment.

The consequence of these demands is that the indicator of strong sustainability requires all natural capital to be measured in physical units.

For example, Dasgupta and Mäler (2001) have argued that prices can fully reflect sustainability and the limits to substitution. Hamilton (2000) points out the restrictive and unlikely conditions that must be fulfilled in order for prices to provide a true measure of sustainability.

To this purpose, Hueting's Sustainable National Income (SNI) is a methodology which estimates what the level of national income would be if the economy met all environmental standards using currently available technology. Hueting's SNI is the maximum income that can be sustained without technological development (excluding the use of non-renewable resources). It is not meant to represent what the economy should look like, but rather to show to policy-makers and the public the distance between the current economy and a sustainable economy. Because of the complexity involved in calculating an SNI, no studies have produced comparable SNIs across countries.

In the following, we take only two different approaches into account, the NAMEA and SNI, as the most representative examples of integrated environmental and economic accounts, highlighting the advantages and disadvantages of both approaches.

The NAMEA

Here we would like to give a short description of the NAMEA as it was developed in the Netherlands. We abstain from explaining the details and how the numbers in the NAMEA are calculated. We only want to give a brief overview of the NAMEA which explains what kind of information the NAMEA can provide for policy-makers.

The NAMEA is a statistical information system which combines national accounts and environmental accounts in a single matrix. The NAMEA is a satellite accounting matrix (SAM), as it is described in the SNA 1993 (chapter 21).[6] The concept of the NAMEA system is based on the work of Keuning (1992, 1993), de Haan and Keuning (1996) and de Boo *et al.* (1993). The origin of their work is the input–output approach[7] of Leontief (1970).[8] The NAMEA is a so-called satellite and can be added to a usual input–output table for analyses (see Chapter 8 in this book).

The NAMEA system itself contains no economic assumptions; it is only descriptive. It maintains a strict borderline between economic and environmental aspects. It is represented in monetary units on the one hand and in physical units on the other and, for this reason, it is called a hybrid accounting system.

To obtain a clear understanding of the inter-relationships between the natural environment and the economy, we must use their physical representation. If not, we are unable to understand these relations. If the NAMEA system were to assign monetary values to environmental problems, two problems would occur. First, the environment must be valued in monetary units. Second, differentiating between price changes and quantity changes is very complicated. Here we will give an idea of the structure of a NAMEA in order to explain the concept.

An extremely simplified version of the Dutch NAMEA is given in Table 2.1 for 2006.[9] Here we take only two economic variables into account: the GDP and employment, and two environmental variables: greenhouse gases (GHG) and waste. Of course, these accounts have, in fact, far more variables. To give a general insight, information can be obtained from Table 2.1. Additionally, we have reduced all economic world to the agricultural sector, industrial sector, service sector and consumption.[10]

These numbers obviously do not give us much information and for this reason, it makes sense to express the values in percentages, as in Table 2.2. For example, we can see from the first row that agriculture accounts for 2.5 per cent of the GDP and 12 per cent of GHG.

Table 2.1 A simplified version of the Dutch NAMEA

Economic sectors	*Economic indicators*		*Environmental indicators*	
	GDP (mill. EUR)	*Employment (1,000 labour years)*	*GHG (mill. kg)*	*Waste (mill. kg)*
Agriculture	25,831	211.3	27,996	3,999
Industry	382,702	1,332.3	110,811	41,317
Services	613,250	5,039.3	59,790	7,296
Households	0	0	37,336	9,417
Aggregate	1,021,783	6,582.9	235,933	62,029

Source: data CBS and own calculations.

To make it more obvious, we can reformulate the tables to the relative contributions to the variables. What Table 2.3 tells us is that agriculture contributes 4.69 more times GHG to the environment per unit of GDP. In this respect, the agricultural sector is the most polluting sector in the Netherlands.

The fundamental idea of the NAMEA is to extend the conventional national accounting matrix with two additional accounts. The first additional account is the account for environmental problems such as the greenhouse effect or ozone layer depletion.[11] The selected environmental themes are partly global environmental problems and partly national and local environmental problems. For example, greenhouse gases are globally relevant, whereas acidification of the soil is mostly locally relevant. The local themes can differ from region to region. The second additional account represents environmental substances such as carbon dioxide or sulphur dioxide where these substances are expressed in physical quantities.[12]

The selection of themes and substances should follow environmental problems considered by policy-makers in the region as the most important. In the Netherlands that is done by the Netherlands Ministry of Housing, Spatial Planning and the Environment (1989, 1990, 1992, 1993)[13] and approved by the Dutch parliament. We can see that the NAMEA generates consistent summary indicators for the environmental problems which are considered to be most pressing at political level.

Table 2.2 A simplified version of the Dutch NAMEA (%)

Economic sectors	*Economic indicators*		*Environmental indicators*	
	GDP	*Employment*	*GHG*	*Waste*
Agriculture	2.5	3.2	12	6
Industry	37.5	20.3	47	67
Services	60	76.5	25	12
Households	0	0	16	15
Aggregate	100	100	100	100

Source: own calculations.

Table 2.3 GHG to GDP sector ratio

	Relative GHG to GDP figures
Agriculture	4.69
Industry	1.25
Services	0.42
Aggregate	1.00

Source: own calculations.

In a certain sense, the NAMEA tables show the boundaries of the core national accounts. The physical accounts of the NAMEA expand these boundaries. The NAMEA distinguishes between households and industries, including public services as well.

Since the compilation of the NAMEA is explained in Keuning (1992) and Keuning *et al.* (1999), we will refer to them in detail.[14]

If NAMEA tables are available for different periods, we can also see how the profiles of economic activities change from time to time. These aspects are highly relevant to policy-makers and future estimations.

For example, de Boer *et al.* (1994) make use of a model with the data from the NAMEA to estimate the consequences of reducing the pollution levels to norms set by the Dutch parliament. Verbruggen *et al.* (1996) make estimations for different scenarios concerning sustainable economic development of the Netherlands until 2030. Without any doubt, the results of these model estimations hinge on assumptions made with regard to the behaviour of the rest of the world and assumptions regarding technical progress made to improve eco-efficiency.

If we consider these tables for certain periods, as in de Haan and Keuning (1995, 1996) or Keuning and de Haan (1997), we can break down the changes in emissions by industry into several effects:[15] demand composition shift effects, output growth effects and eco-efficiency change effects. The first effect can be positive or negative in the sense that the claims to use the natural environment are reduced. The second effect is negative because more output generally means an increased use of the natural environment because of the laws of thermodynamics. The third effect is positive because of technological progress. De Haan (1996), for example, has connected the NAMEA with data on estimated costs and emission reductions in a range of potential energy-saving measures by industry in the Netherlands. He came to the conclusion that the Dutch economy would be better off to some extent if the most efficiency measures were applied first. However, if the norms for CO_2 emissions set by the government were too restrictive, the result would be the reverse.

The NAMEA is a multi-purpose information system which is able to inform the public and policy-makers of the status quo of environmental assets and environmental pollution. More specifically, NAMEA provides policy-makers with a data framework which can be used to sketch the trade-off between the prevention of environmental damage and macro-economic policy objectives. Selecting which kind of environmental problem should be represented depends on political decisions and not on scientists' decisions. This is the reason why NAMEAs in different countries are different. The British NAMEA, for example, contains 15 environmental substances and only three environmental themes (Vaze, 1999), the Japanese one contains 16 substances and six themes (Ike, 1999), the German one contains eight substances and two themes (Tjahjadi *et al.*, 1999) and the Swedish NAMEA contains five substances (Hellsten *et al.*, 1999).[16] Without a doubt, it would be useful to standardize the NAMEAs in all countries, because of global environmental issues. It is nevertheless worth noting that Eurostat is releasing an EU27 NAMEA by 2011 covering 2000–2006.

Policy issues can be tackled. The data from the NAMEA can be used to calculate the effects of a shift in tax incidence, from labour to energy use, for example, on environmental and economic indicators in the NAMEA system. Additionally, the data can be used to model a general equilibrium model for estimating the consequences of a change in the tax system. With the help of the NAMEA, the consequences of specific political decisions can be calculated. For example, the reduction of acid substances caused by the introduction of catalytic converters in cars can be assessed. In the Netherlands, this resulted in a decrease in the burden of ozone layer depletion by nearly 12.3 per cent. In addition, new insights can be gained into the issues of who should pay for the environmental damage.[17]

As a result, the NAMEA is not only used to derive aggregate indicators from a consistent meso-level information system, it also provides data in the required format for all kinds of analyses.

The sustainable national income of Roefie Hueting

The aim of this section is to show what kinds of problems are connected with the concept of GNI. Since there are a number of distinct approaches and definitions of a GNI in the literature,[18] we will concentrate on the concept of SNI as defined by Hueting (1974, 1980). There are two specific reasons why we investigate SNI. The first reason is that the SNI concept was used by Dutch econometricians who wanted to calculate the differences between the actual NNI in the Netherlands and Hueting's SNI (1974, 1980). The second reason is that some Dutch economists and policy-makers have proposed introducing the concept of the SNI into the official statistics of the Netherlands. However, before we start with the analysis, we will explain Hueting's ideas. More specifically, we will look at Hueting's most relevant assumptions and how he wants to justify these assumptions.

According to Hueting, the system of national accounts (SNA) should be corrected for environmental losses or rather the monetary cost to abate these losses. If not, he fears that some important welfare losses of an economy are ignored.[19] He calls, in particular, for the introduction of a practical concept of sustainability into the national accounting system. Hueting's contributions concern the relationship of the indicators for the NNI and the SNI. It is important to see that his work is founded theoretically and then applied to economic statistics. His objective is to provide adequate information to the users of statistical data on the state of the natural environment. This section is mainly based on the work of Hueting (1974, 1980), Hueting and de Boer (2001), Hueting and Reijnders (1998) and Hueting *et al.* (1992, 1995). Hueting assumes that we (the inhabitants of the world) prefer the complete conservation of our natural environment to reach strong sustainability.[20] The foundation for his view on sustainability goes back to J.S. Mill's (1876) concept of *steady state* and *stationary state*.[21] This implies that it is inadmissible to transfer environmental risks and burdens to future generations. The natural environment must be conserved by the living generation. This consideration is based on the principle of preferences for intergenerational equity. Hueting's idea is to calculate the costs for the conservation of the natural environment and to subtract these costs from the NNI. To establish an appropriate maximum environmental burden to meet these preferences, it is seen as a task for natural scientists. Formally, the SNI is defined as:

$$SNI = NNI - \sum_{i=1}^{N} C(D_i), \tag{1}$$

where N is the number of harmful substances such as CO_2, SO_4 and so on, D_i is the amount of substance i which is emitted above a specific sustainability level of this substance and C represents the avoidance cost function for this substance. Given these assumptions, it follows that the value of environmental degradation is equal to the conservation costs.[22] Additionally, given that these costs are known, an SNI can be calculated. This is the difference between the NNI minus the aggregated costs to preserve the natural environment from degradation. Or in the words of Hueting and de Boer:

> The SNI according to Hueting is the maximum net income which can be sustained on a geological time scale, with future technology progress assumed only in the development of substitutes for non-renewable resources, where such substitution is indispensable for sustaining environmental functions, in turn essential for sustaining income.
>
> (2001, pp. 19 and 70)

The gap between the NNI and SNI measures the dependence of the economy on its natural environment. If the gap increases, the economy becomes more unsustainable. If the gap decreases, the economy becomes more sustainable. Hueting's assumptions avoid the problem that we must have knowledge about the future. Otherwise, we will run into unsolvable problems.[23] Therefore, assumptions about preferences must be made. There are different reasons to assume that objective strong sustainability can be justified. First, Hueting argues that most countries have agreed that reaching strong sustainability is a policy objective since the publication of the Brundtland report.

However, there are some arguments in the literature to reject the SNI as a statistical tool. For example, if the use of environmental resources is prevented, the structure of the 'sustainable' economy will differ from the actual structure of the economy. Second, it is unclear what will happen with the international trade structure. It would be a strong mistake to ignore these changes. Consequently, a theoretical model must be used which considers the whole economy, to calculate Hueting's SNI. As a result, we get a hypothetical SNI.[24] To calculate this hypothetical SNI for the Netherlands, Verbruggen *et al.* (2001) have made some additional assumptions:[25] (a) the individual preferences for the sustainable use of the environment are absolute and independent of costs. This implies that the aggregate demand curve for environmental functions is absolute price-inelastic; (b) the instantaneous realization of sustainability standards and no transition costs, implying that Hueting's approach is a static one; (c) the sustainability standards are applied all over the world to avoid arbitrage effects between different countries; (d) labour market effects are ignored; no change in the unemployment rate is assumed; (e) the SNI must be calculated with sustainable relative prices; (f) the government does not change its politics in the sustainable economy.

Gerlagh *et al.* (2001, 2002) and Dellink *et al.* (2001) combine Hueting's SNI with an AGE (Applied General Equilibrium) model to calculate the SNI for the Netherlands in 1990. To model the natural environment they use data from Keuning's (1993) NAMEA and from de Boer (2002). However, the results of the various scenarios are different, but in principle it can be concluded that one half of the Dutch GDP was produced in an unsustainable way. Additionally, the reader might ask if the assumptions are realistic and if the results can be used for policy advising and policy-making. If any government implements a policy to conserve the natural environment and the national income is substantially cut, the results will be dangerous for the stability of the society if the protection costs are financed in a way that changes income distribution. This is because it is likely that the whole economy must be totally changed to obtain sustainability. A 'transition' approach to environmental policy and sustainability is then necessary.

The next point is that from politicians' viewpoint, it is very risky to assume that there will be no technical progress in environmental technologies which would lower conservation or reconstructing costs. From Thomas Malthus to Stanley Jevons and the Club of Rome, predictions of the imminent ruin of humankind have so far been totally wrong (Beckerman, 2001).

In light of these problems, the Scientific Council for Government Policy of the Netherlands (WRR) states:

> The fact that in abstracto there are scientific limiting conditions on behaviour would appear clear enough.... It is however an entirely different matter to determine in concreto whether those limits have been reached or are possibly already being breached, or whether they will come into effect at a point far beyond the relevant time-horizon for decision-making.

Further on: 'The available knowledge is very much fragmentary in nature and the (dynamic) interactions between various sub-elements of the "system Earth" go beyond the human capacity for understanding' (WRR, 2002, p. 19).

Keuning (1992) argues against the SNI as an accounting tool; he especially criticizes how the SNI is measured when he asserts: 'Replacing the GDP (Gross National Product, P.S.) by a figure which is an erratic combination of statistic and the outcome of an (implicit) model thus amounts to throwing out the baby with the bath-water.'

El Serafy (1997, p. 221) also criticizes the concepts of 'green accounting' in general:

> When current prices are used for (environmental) stock valuation, and changes in stock values are incorporated in the flow accounts, the integrity of the latter is damaged, and very little environmental wisdom will be gained from such procedure, and even less economic insight.

Further on he also argues that 'accounting in physical units, or in indices based on physical units, are best for revealing environmental change' (El Serafy, 1997, p. 224).

Finally, there are some fundamental critics from Norgaard *et al.* (1998) and others who argue that is impossible to fully value the environment in monetary units.

Conclusions

Now let us summarize what we have learned from the investigation into the NAMEA and SNI approach. First, it should be noted that the NAMEA system is really only a descriptive statistical tool where no economic assumptions should be made. As a result of this, the NAMEA in its naked formulation does not contain an implicit or explicit policy implication.

The most important argument for the NAMEA approach is probably that the accounting system, though it is a 'heterodox' tool, can be used by various economic perspectives and disciplines. The main reason is that it is free of value statements. The SNI approach, on the other hand, was rejected as a national accounting tool by many national accountants. The reason for this is that the SNI is based on a number of critical assumptions and implicitly, the SNI is based on the 'ideology' of strong sustainability. That means that policy-makers do not decide on environmental and implicitly on optimal economic policy, but the SNI forces a specific policy. For this reason, the SNI was for a long time preferred in Dutch policy and among the public because it delivers only one main indicator which is easy to understand. The NAMEA, on the other hand, is able to deliver an endless number of different indicators, depending on the number of economic and environmental indicators it aims at and needs to analyse. Much discussion has taken place in the scientific community and in the Dutch parliament over a period of 15 years. Now, the fact that the NAMEA is the 'strongest tool' for supporting the promotion of well-being and progress, as stated by commissioner Dimas in 2007, has been decided at European level.[26] Compiling a NAMEA for all member states of the European Union was laid down in the European Strategy for Environmental Accounting (ESEA) in 2003. Additionally, the NAMEA is part of the System of Environmental and Economic Accounts (SEEA) as described in UN (2003). In Europe, only Luxembourg and Malta do not compile environmental data and Greece and Slovakia only compile environmental expenditure data. All other European countries follow the 2003 and 2008 guidelines of the ESEA. The countries are supported by the European Environment Agency.

It should also be noted that the introduction of the NAMEA in many countries was only possible because of the political independence of statistical offices in most countries, especially in Europe. Without this independence, a green national income would probably have been introduced because policy-makers prefer striking and simple indicators. Of course, it took more than 20 years to find a compromise in science, but it worked.[27]

Notes

1 See Aaheim and Nyborg (1995) or Lange (2003) for an overview.
2 Lange (2003).
3 A more detailed analysis of the concept of sustainability is given by Ayres *et al.* (1998), Gowdy (2004) and Heal (1996).
4 Until now, human capital has not been included in the official national accounts because there is no agreement on how to measure it (some researchers have tried to do this, see Stauvermann (1997) and the cited literature there).
5 See Hartwick (1977) and Solow (1986).
6 The original idea of the SAMs was to incorporate concerns of inequality and poverty in the national accounts and input–output tables. An introduction to the SAM approach is given in Keuning and de Ruijter (1988), Pyatt and Thorbecke (1976), Pyatt and Round (1986) and Alarcon *et al.* (1991).
7 Duchin and Steenge (1999) give a brief technical overview of input–output analysis with respect to environmental problems.
8 Leontief's (1970) analysis of the physical economy can be regarded as the first prototype NAMEA since both systems are characterized by a hybrid structure including both physical as well monetary data (de Haan, 2001, p. 5).
9 See the CBS web page: www.statline.cbs.nl.
10 Of course, most of the sectors are represented in much more detail and follow the Standard Industrial Classification for the most part.
11 The numbers for the environmental themes are aggregated with the help of IPCC conventions. This means, for example, that 1 kg CO_2 emissions equals one global warming potential, 1 kg N_2O emissions equals 270 global warming potentials and 1 kg CH_4 equals 11 global warming potentials.
12 See, for example, the NAMEA table in Keuning *et al.* (1999, pp. 18–22).
13 The pilot NAMEA in 1993 benefited a great deal from work done on environmental indicators at the Ministry of Housing, Spatial Planning and the Environment (Adriaanse, 1993).
14 The NAMEA 1995 is given in the appendix of Keuning *et al.* (1999).
15 An explanation of how to do this is given in Kee and de Haan (2004) and de Haan (2000).
16 For a comparison of the different approaches, see de Haan (1999).

17 See Steenge (1997, 1999).
18 See Adriaanse (1993) and Peskin (1998).
19 Hueting has published more than 75 articles, papers and books in English on this topic. Goodland (2001, pp. 326–331) gives an overview of his work.
20 In the literature (Goodland, 1995) there are several different definitions of sustainability: e.g. weak sustainability, strong sustainability. Hueting defines sustainability as a situation in which vital environmental functions remain available *ad infinitum*. In Hueting's view, sustainability can be defined as being scientifically objective (Hueting and Reijnders, 1998; Reijnders, 1996).
21 Steady state and stationary state means a state which is sustainable forever. For details, see Stauvermann (1997).
22 The costs contain the costs of preserving the environment and the costs of removing existing environmental burdens.
23 For example, the models of Weitzman (1976) and of Hartwick (1977), which was built on Solow (1974), are based on very strong assumptions: identical consumer preferences, certain future, no technical change, constant time preferences of the consumers and no distortionary taxes or subsidies. The results of these models break down if we relax these assumptions.
24 See Verbruggen *et al.* (2001, p. 277). We then envisage a hypothetical sustainable economy with a hypothetical SNI.
25 This study was overseen by a committee chaired by Frank den Butter and was financed by the Dutch ministries of Hans Wijers and Margreeth de Boer.
26 See www.beyond-gdp.eu.
27 See Geerdink and Stauvermann (2009).

References

Aaheim, A. and Nyborg, K. (1995) 'On the interpretation and applicability of a green national product', *Review of Income and Wealth*, 41: 57–71.

Adriaanse, A. (1993) *Environmental Policy Performance Indicators*, Netherlands Ministry of Housing, Spatial Planning and the Environment.

Alarcon, J.V., van Heemst, J., Keuning S.J., de Ruijter, W.A. and Vos, R. (1991) *The Social Accounting Framework for Development: Concepts, Construction and Applications*, Aldershot: Avebury.

Ayres, R.U., van den Bergh, J.C.J.M. and Gowdy, J.M. (1998) 'Viewpoint: weak versus strong sustainability', discussion paper no. 98-103/3, Amsterdam: Tinbergen Institute.

Beckerman, W. (2001) 'Technical progress, finite resources and intergenerational justice', in E.C. van Ierland, J. van der Straaten and H.R.J. Vollebergh (eds), *Economic Growth and Valuation of the Environment: A Debate*, London: Edward Elgar.

Brekke, K.A. (1997) 'Hicksian income from resource extraction in an open economy', *Land Economics*, 73: 516–527.

Common, M. and Perrings, C. (1992) 'Towards an ecological economics of sustainability', *Ecological Economics*, 7: 7–34.

Dasgupta, P. and Mäler, K.G. (2001) 'Wealth as a criterion for sustainable development', *World Economics*, 2: 19–44.

De Boer, B. (2002) *Economische Structurenen milieugebruik: basisrapport van het project Duurzaam Nationaal Inkomen, concept*, Voorburg: Statistics Netherlands.

De Boer, B., de Haan, M. and Voogt, M. (1994) 'What would Net Domestic Product have been in a sustainable economy? Preliminary views and results', in Statistics Canada (ed.), *National Accounts and the Environment: Papers and Proceedings from a Conference*, London, Ottawa, 16–18 March.

De Boo, A., Bosch, P., Gorter, C.N. and Keuning, S.J. (1993) 'An environmental module and the complete system of national accounts', in A. Franz and C. Stahmer (eds), *Approaches to Environmental Accounting*, Heidelberg: Physica Verlag.

de Haan, M. (1999) 'On the international harmonisation of environmental accounting: comparing the National Accounting Matrix including Environmental Accounts of Sweden, Germany, the UK, Japan and the Netherlands', *Structural Change and Economic Dynamics*, 10: 151–160.

de Haan, M. (1996) 'An input–output estimation of avoidance costs', paper presented at the 4th Biennial Meeting of the International Society for Ecological Economics, Boston, 4–7 August.

de Haan, M. (2000) 'A structural decomposition analysis of pollution in the Netherlands', *Economic Systems Research*, 13: 181–196.

de Haan, M. (2001) 'Physical macroeconomics: a demarcation of accounting and analysis', paper presented at the Economic Growth Material Flows and Environmental Pressure Workshop, Stockholm, 26–27 April.

de Haan, M. (2004) *Accounting for Goods and for Bads: Measuring Environmental Pressure in a National Accounts Framework*, Voorburg: Statistics Netherlands.

de Haan, M. and Keuning, S.J. (1995) 'Taking the environment into account: the Netherlands NAMEAs for 1989, 1990 and 1991', Occasional Papers NA-074, Voorburg: Statistics Netherlands.

de Haan, M. and Keuning, S.J. (1996) 'Taking the environment into account: the NAMEA approach', *Review of Income and Wealth*, 42: 131–148.

de Haan, M., Keuning, S.J. and Bosch, P.R. (1994) 'Integrating indicators in a National Accounting Matrix including Environmental Accounts (NAMEA): an application to the Netherlands', in Statistics Canada (ed.), *National Accounts and the Environment: Papers and Proceedings from a Conference*, London, Ottawa, 16–18 March.

Dellink, R., Gerlagh, R. and Hofkes, M.W. (2001) *Calibration of an Applied General Equilibrium Model for the Netherlands 1990*, IVM Report W-01/17, Amsterdam: Institute for Environmental Studies.

Duchin, F. and Steenge, A.E. (1999) 'Input–output analysis, technology and the environment', in J.C.J.M. van den Bergh (ed.), *Handbook of Environmental and Resource Economics*, Northampton, MA: Edward Elgar.

El Serafy, S. (1997) 'Green accounting and economic policy', *Ecological Economics*, 21: 217–229.

Geerdink, G.C. and Stauvermann, P.J. (2009) 'A pleading for policy-independent institutional organizations', in A. Prinz, G.J. Hospers and B. Bunger (eds), *Squaring the Circle: Essays in Honor of Bert Steenge*, Munster: LIT-publishers.

Gerlagh, R., Dellink, R., Hofkes, M. and Verbruggen, H. (2001) *Calibration of an Applied General Equilibrium Model for the Netherlands*, IVM Report W-01/16, Amsterdam: Institute for Environmental Studies.

Gerlagh, R., Dellink, R., Hofkes, M. and Verbruggen, H. (2002) 'A measure of sustainable national income for the Netherlands', *Ecological Economics*, 41: 157–174.

Goodland, R. (1995) 'The concept of environmental sustainability', *Annual Review of Ecological Systems*, 26: 1–24.

Goodland, R. (2001) 'An appreciation of Dr Roefie Hueting's ecological work', in E.C. van Ierland, J. van der Straaten and H.R.J. Vollebergh (eds), *Economic Growth and Valuation of the Environment: A Debate*, London: Edward Elgar.

Gowdy, J.M. (2004) 'Toward a new welfare foundation for sustainability', Working Papers in Economics 0401, Renesselaer Polytechnic Institute.

Hamilton, K. (2000) 'Genuine savings as a sustainability standard', Environmental Economics Report No. 77, Washington, DC: World Bank.

Hartwick, J.M. (1977) 'Intergenerational equity and the investing of rents from exhaustible resources', *American Economic Review*, 67: 972–974.

Hartwick, J.M. (1990) 'Natural resources, national accounting and economic depreciation', *Journal of Public Economics*, 43: 291–304.

Heal, G.M. (1996) 'Interpreting sustainability', Columbia University, mimeo.

Hellsten, E., Ribacke, S. and Wickbom, G. (1999) 'SWEEA: Swedish environmental and economic accounts', *Structural Change and Economic Dynamics*, 10: 39–72.

Hicks, J.R. (1946) *Value and Capital*, Oxford: Clarendon Press.

Hofkes, M.W., Gerlagh, R., Lise, W. and Verbruggen, H. (2002) 'Is economic growth sustainable? A trend analysis for the Netherlands, 1990–1995', discussion paper, IVM Institute for Environmental Studies, Amsterdam: Vrije Universiteit.

Hueting, R. (1974) *Nieuwe schaarste en economische groei*, Amsterdam: North-Holland.

Hueting, R. (1980) *New Scarcity and Economic Growth: More Welfare through Less Production?* Amsterdam: North Holland Publishing Company [trans. of Hueting, 1974].

Hueting, R. and de Boer, B. (2001) 'The parable of the carpenter', *International Journal of Environment and Pollution*, 15: 42–50.

Hueting, R. and Reijnders, L. (1998) 'Sustainability is an objective concept', *Ecologic Economics*, 27: 139–147.

Hueting, R., Bosch, P. and de Boer, B. (1992) 'Methodology for the calculation of sustainable national income', Statistical Essays M44, Voorburg: Netherlands Central Statistical Office.

Hueting, R., Bosch, P. and de Boer, B. (1995) 'The calculation of sustainable national income', Occasional Papers and Reprints, IDPAD OPand R 1995–2, Indo-Dutch Programs on Alternatives in Development IDPAD, New Delhi, The Hague.

Ike, T. (1999) 'A Japanese NAMEA', *Structural Change and Economic Dynamics*, 10: 123–149.

Kee, P. and de Haan, M. (2004) 'Accounting for sustainable development', discussion paper, Voorburg: CBS.

Keuning, S.J. (1991) 'A proposal for a SAM which fits into the next system of national accounts', *Economic Systems Research*, 3: 233–248.

Keuning, S.J. (1992) 'National accounts and the environment: the case for a system's approach', Occasional Paper NA-053, Voorburg: Statistics Netherlands (CBS).

Keuning, S.J. (1993) 'An information system for environmental indicators in relation to the national accounts', in W.F.M. de Vries, G.P. den Bakker, M.B.G. Gircour, S.J. Keuning and A. Lenson (eds), *The Value Added of National Accounting*, Voorburg: Statistics Netherlands.

Keuning, S.J. and de Haan, M. (1997) 'Netherlands: what's in a NAMEA? Recent results', in K. Uno, P. Bartelmus and C. Stahmer (eds), *Environmental Accounting in Theory and Practice*, Dordrecht: Kluwer.

Keuning, S.J. and de Ruijter, W.A. (1988) 'Guidelines to the construction of a national accounting matrix', *Review of Income and Wealth*, 34: 71–101.

Keuning, S.J. and Steenge, A.E. (1999) 'Introduction to the special issue on "Environmental extensions of national accounts: the NAMEA framework" ', *Structural Change and Economic Dynamics*, 10: 1–13.

Keuning, S.J., van Dalen, J. and Haan, de M. (1999) 'The Netherlands' NAMEA: presentation, usage and future extensions', *Structural Change and Economic Dynamics*, 10: 15–37.

Lange, G.M. (2003) 'Policy applications of environmental accounting', Environmental Economics Series, World Bank Environment Department, Washington, DC: World Bank.

Leontief, W. (1970) 'Environmental repercussions and the economic structure: an input–output analysis', *Review of Economics and Statistics*, 52: 262–271.

Mäler, K.G. (1991) 'National accounts and environmental resources', *Environmental and Resource Economics*, 1: 1–15.

Norgaard, R.B., Bode, C. and Values Reading Group (1998) 'Next, the value of God, and other reactions', *Ecological Economics*, 25: 37–39.

Peskin, H.M. (1998) 'Alternative resource and environmental accounting approaches', in K. Uno and P. Bartelmus (eds), *Environmental Accounting in Theory and Practice*, Dordrecht: Kluwer.

Pyatt, G. and Round, J. (1986) *Social Accounting Matrices: A Basis for Planning*, Washington, DC: World Bank.

Pyatt, G. and Thorbecke, E. (1976) *Planning Techniques for a Better Future*, Geneva: International Labor Office.

Reijnders, L. (1996) *Environmentally Improved Production Processes and Products*, Dordrecht: Kluwer.

Solow, R.M. (1974) 'Intergenerational equity and exhaustible resources', *Review of Economic Studies*, Symposium on the Economics of Exhaustible Resources: 29–45.

Solow, R.M. (1986) 'On the intergenerational allocation of natural resources', *Scandinavian Journal of Economics*, 88: 141–149.

Stauvermann, P. (1997) 'Endogenous growth in OLG-models', Wiesbaden, mimeo.

Steenge, A.E. (1997) 'On background principles for environmental policy: polluter pays, user pays or victim pays', in B. Boorsma, K. Aarts and A.E. Steenge (eds), *Public Priority Setting: Rules and Costs*, Dordrecht: Kluwer.

Steenge, A.E. (1999) 'Input–output theory and institutional aspects of environmental policy', *Structural Change and Economic Dynamics*, 10: 161–176.

Tjahjadi, D., Schaefer, D., Radermacher, W. and Hoeh, H. (1999) 'Material and energy flow accounting in Germany: data base for applying the National Accounting Matrix including Environmental Accounts concept', *Structural Change and Economic Dynamics*, 10: 73–97.

Vaze, P. (1999) 'A NAMEA for the UK', *Structural Change and Economic Dynamics*, 10: 99–121.

Verbruggen, H., Bennis, M.J., Dellink, R.B., Jansen, H.M.A., Kuik, O.J. and Ruygrok, E.C.M. (1996) 'Duurzame Economische Ontwickkelings-Scenario's voor Nederland in 2030', Publikatiereeks milieustrategie, No. 1996/1, The Hague: Ministrie van Milieu, Volkshuisvesting, Ruimtelijke Ordening an Milieu (VROM).

Verbruggen, H., Dellink, R.B., Gerlagh, R., Hofkes, M.W. and Jansen, H.M.A. (2001) 'Alternative calculations of a sustainable national income for the Netherlands according to Hueting', in E.C. van Ierland, J. van der Straaten and H.R.J. Vollebergh (eds), *Economic Growth and Valuation of the Environment: A Debate*, London: Edward Elgar.

Weitzman, M.L. (1976) 'On the welfare significance of national product in a dynamic economy', *Quarterly Journal of Economics*, 90: 156–162.

WRR (2002) *Sustainable Development: Administrative Conditions for an Activating Policy*, Reports to the Government 62, The Hague.

3 Air emissions in Italian regions

The role of technological and geographical spillovers

Valeria Costantini, Massimiliano Mazzanti and Anna Montini

Introduction

This chapter investigates the economic drivers which may influence the geographical distribution of environmental performance by using a new and innovative hybrid environmental-economic accounting matrix applied to the Italian regions, based on the NAMEA (National Accounting Matrix including Environmental Accounts). On a national basis, it was mostly developed in the late 1980s and 1990s in the Netherlands[1] to respond to the awareness of the Dutch public and policy-makers of environmental problems (see Stauvermann in Chapter 2). In Italy, the Italian National Statistical Institute (ISTAT) has been regularly releasing a time series of air emission accounts at national level since 2004; regionalization of data generation has led to an Italian regional NAMEA for the year 2005, recently published by ISTAT (2009), involving 20 regions, 24 productive sectors and 10 pollutants which results in a quite extensive and unique dataset (see Tudini and Vetrella in Chapter 1 for the analytical merging process of the economic (NAM) and environmental (EA) components). The great advantage of this new tool is that it adds a geographical dimension to the sectoral one which already exists, disentangles the structural and efficiency factors behind a regional environmental performance and assesses which drivers are relevant to determining the distribution of environmental performance at regional and sectoral level.

As emphasized by Gibbs (2006), regional analyses based on economic and environmental accounts may contribute to establishing grounds for fruitful research in environmental issues that remain comparatively under-researched and provide normative prescriptions for a properly designed environmental policy in a context of geographical and economic heterogeneity.

While an increasing number of empirical analyses emphasized the potential role of a sector-based investigation in describing the environmental efficiency patterns of distinguished economic sectors (de Haan, 2004; de Haan and Keuning, 1996; Femia and Panfili, 2005; Mazzanti and Montini, 2010a, 2010b; Mazzanti *et al.*, 2008), most of the analyses lack an explicit consideration of how the technological[2] and composition effects are embedded in a regional context. To the best of our knowledge, there are no attempts to investigate which kind of innovation prevails when shaping the delinking process.[3] To this end, we argue that internal innovation partially explains environmental efficiency gains whereas the role of knowledge spillovers may help to discover if the technology diffusion process will improve environmental performance as well as economic growth.

In this context, recent efforts in the economic geography literature will give us useful analytical tools for shaping the role of innovation and spillovers at regional level which is currently developed mainly for discovering impacts on economic growth. Bearing in mind that economic growth and productivity gains have been addressed as driving factors explaining environmental performance, we may well adapt an analytical framework of this type to our purpose.

In particular, we refer to the extensive literature debating the impact of different kinds of agglomeration economies on technological innovation patterns and economic development at regional level. More specifically, what we are interested in is discovering if the well-known existing agglomeration effects in economic terms for the Italian regions may also explain the geographical distribution of environmental performance. If a clustering process occurs and environment-friendly or 'hot spot' areas emerge, we argue that some forms of spillovers between regions and sectors may help us explain environmental performance better than using only the traditional driving forces proposed by the environmental economics literature.

This specific assumption could be taken in a sector-based analysis when regional features are also accounted for. As we will specify more clearly in the model setting, when sector-specific and regional fixed effects are included, a negative correlation between technological innovation and environmental inefficiency may well be interpreted as the positive role of the general innovative capacity on environmental achievements. This is particularly recommended when strong agglomerative effects emerge in a descriptive analysis. Since Italian regional manufacturing sectors are historically characterized by clusters and agglomeration economies (Cefis *et al.*, 2009), the role of the centripetal and centrifugal forces are also assumed to be crucial to explaining environmental performance.

The original contribution of this chapter is to explore how environmental efficiency is distributed among regions and sectors in the Italian context and trying to discover if agglomerative effects occur and if they correspond to a regional or sectoral criterion. When the geographical distribution of environmental performance is characterized by a clustering effect, we also present a methodological proposal in order to underline which driving factors influence such agglomerative phenomena, with a particular emphasis on the role of geographical spillovers both on the environmental and the technological side.

Analysing the geographical distribution of environmental performance

Methods

This section provides a conceptual background for the empirical analyses we have adopted to describe the geographical distribution of environmental performance at regional level and investigate the relative role of key driving factors that influence this type of distributive pattern. For the first descriptive purpose, we choose a decomposition approach, represented by shift-share analysis, in order to catch whether a region-based or a sector-based criterion prevails in the allocation of different environmental performance whereas for the second one, we propose an econometric estimation. We emphasize the fact that these two analytical tools pursue different but complementary aims. The former gives a preliminary sketch of regional environmental performance features where sector-based clustering effects seem to exist independently of geographical patterns, thus revealing the need for a better understanding of which factors influence such agglomerative effects the most. To explore the role of regional productive structures in emission efficiency across regions, shift-share analysis (Esteban, 1972, 2000) is thus used to decompose the source of change of the specified dependent variable into regional specific components (the *shift*) and the portion that follows national growth trends (the *share*). Our starting point in this study is the aggregate indicator of emission intensity, represented by total emissions of a particular pollutant on value added, defined as (E/Y) for Italy as the benchmark, and as (E^r/Y^r) for the analysed *r-th* region. This indicator is decomposed[4] as the sum of $(E_k/Y_k)*(Y_k/Y)$ where (Y_k/Y) is the share of sectoral value added on total value added, for the *k-th* sector, with k defined from one to n (where n=24 NACE sectors included in the regional NAMEA).[5]

According to preliminary findings of the decomposition approach, the econometric estimation can be used to quantify to which extent sectoral and regional features influence emission efficiency by also considering the role of spatially related externalities as environmental and innovation spillovers. By considering environmental pressure – here expressed through pollutant emissions – for each *k-th* sector in each *r-th* region (E_k^r) as a function of production level (Y_k^r), technology (T_k^r) and environmental policy (P_k^r) as suggested by Cole *et al.* (2005), emissions can be expressed as the following general function:

$$E_k^r = f\left(Y_k^r, T_k^r, P_k^r\right) \tag{1}$$

Since we are interested in an evaluation of the environmental performance of our sector expressed as a measure of emission intensity, we can transform the previous equation by scaling it with region/sector-specific value added, thus obtaining the following reduced form:

$$e_k^r = \alpha_k^r + \beta_1 \ln Y_k^r + \beta_2 t_k^r + \beta_3 p_k^r + \varepsilon_k^r \tag{2}$$

A specific variable representing environmental spillovers from other regions should therefore be included in equation (2). Hence, considering both environmental and innovation spillovers, equation (2) is transformed as follows:

$$e_k^r = \alpha_k^r + \underset{-}{\beta_1} \, lp_k^r + \underset{+}{\beta_2} \, es_k^r + \underset{-}{\beta_3} \, t_k^r + \underset{-}{\beta_4} \, ts_k^r + \underset{-}{\beta_5} \, p_k^r + \varepsilon_k^r \tag{3}$$

where lp_k^r is labour productivity, while es_k^r and ts_k^r represent the effects of environmental and innovation spillovers coming from the other Italian regions, empirically modelled as described in Costantini *et al.* (2011). We might expect a positive sign for the β_2 coefficient that depends on the existence of agglomerative forces producing a concentration of dirty activities into circumscribed geographical areas. We affirm that with a properly defined disaggregation of manufacturing activities, environmental regulation and technological innovation strategies may act coherently towards an agglomeration effect of high-tech less-polluting activities. On this basis, we may well expect a positive effect on environmental performance related to stringent environmental policies (p_k^r), or, in other words, in this case the β_3 coefficient is also expected to be negative where the more stringent the regulatory framework is at regional level, the lower the emission intensity is at sectoral level. We also expect β_4 to be negative, coherently with the role played by internal innovation (β_3), since we may well assume that the existence and diffusion of technologies from other regions will increase the probability that a more environment-friendly production technique will be available.

The dataset

The core part of the dataset is based on the 2005 Italian regional NAMEA, to our knowledge the only full *regional* NAMEA available in the EU. Environmental pressures (ten air pollutants) and economic data (value added, household consumption expenditures and full-time equivalent employment) are assigned to the economic branches of resident units. The accounting approach allows a full dataset to be shaped with information on environmental and economic aspects. Our dataset is organized as a $(q \times n) \times 1$ vector where n is the total number of k sectors ($\forall k = 1, \ldots, n$, with $n = 24$) and q is the number of r regions ($\forall r = 1, \ldots, q$, with $q = 20$), with a maximum potential number of observations equal to 480.

In the shift-share analysis, we have considered specific pollutant emissions in order to obtain a clear picture of the distribution at sectoral level of emission intensity among regions since each pollutant may be associated with specific production specialization. When testing the influence of different drivers of environmental performance as expressed by equation (3), we have adopted the environmental theme aggregation tool provided by NAMEA where specific pollutants are summed up as greenhouse gases (GHG) and pollutants responsible for acidification process (ACID).[6] To some extent, this choice enables us to make further considerations on potentially different impacts of the same drivers associated with environmental damage with a different geographical distribution since the effects of GHG are globally distributed whereas ACID emissions are more localized and transboundary effects may be confined to neighbouring regions.

In order to represent the two dimensions of technological innovation, the internal variable (t_k^r) and the inter-regional intra-sector spillover effect (ts_k^r) respectively, we have considered a patent count approach due to more aggregated data available for regional R&D expenditures at sector level. Some drawbacks characterize patents as a valid alternative to R&D data as an economic indicator, but previous studies at regional level have highlighted the helpfulness of patent applications as a measure of production of innovation (Acs *et al.*, 2002).

Patent data are drawn from the REGPAT dataset elaborated by Eurostat from the OECD PATSTAT database, gathering all patents for each region according to the three-digit IPC classification granted by the European Patent Office (EPO), geographically classified by relying on the postal codes of the applicants. The number of patent classes at the three digit-level is 633, and we have considered all patent applications to the EPO by priority year at regional level.

We have adopted an ad hoc sector classification in order to assign patents (as classified by IPC codes) to specific sectors (as classified by NACE codes) relying on previous concordance proposals such as the OECD Technology Concordance and the methodology developed by Schmoch *et al.* (2003), resulting in 13 available sectors (see Table 3.A.2 in the Appendix). As a result of the high variance of patenting activity over time, we have considered patents in the period 2000–2004 in order to calculate a five-year average value as the best proxy of innovation stock at sectoral level with one time lag compared with emissions data (Antonelli *et al.*, in press).

We argue that the potentially positive influence of innovating activities on environmental performance arises with temporal lags since the adoption of new technologies is not exactly simultaneous with the invention itself. Since we are considering the impact of innovation on environmental performance as a side effect of innovative capacity at sectoral level, a one-year lag seems to be the most appropriate choice. Bearing in mind that equation (3) expresses all terms scaled by value added, we have also computed patents to value added ratios in order to account for the innovation intensity of each sector.

In order to include the potential role of inter-regional spillovers, we first consider that the probability of innovation to spill from one region to another strictly depends on the fact that localization economies are associated with the concentration of a particular sector in the two regions. Hence, it is not only a matter of geographical distance which explains the existence and the strength of innovation spillovers, but also cognitive proximity since knowledge is more likely to spread when competences and knowledge stocks of the inventors and adopters are closely related.

Following empirical findings by Costa and Iezzi (2004) on technological spillovers among Italian regions, we have only considered Marshall-type externalities in this chapter (since a sector-specific approach is considered) because innovation spillovers mainly derive from firms belonging to the same industry whereas Jacob-type externalities among sectors are rather smaller. In a certain sense, cognitive proximity and technological relatedness as well-known drivers for effective learning (Boschma and Frenken, 2009; Boschma and Iammarino, 2009, among others) are considered here as factors influencing the adoption of similar production process techniques without any implication in terms of regional economic growth. In our opinion, in this specific context, Marshall-type externalities prevail since the clustering effects of technology-related sectors prevail, because manufacturing sectors are broadly defined here whereas Jacob-type externalities may not be a plausible driver for spillovers. Nonetheless, this last point could be the next research topic, especially when a panel version of the Italian regional NAMEA is available which can be used to consider not only environmental performance but also efficiency increases where dynamic issues are crucial.

Los (2000) and Frenken *et al.* (2007) propose adopting an index that captures the technological relatedness between industrial sectors by computing the similarity between two sectors' input mix from input–output tables that we can adapt to our case study if we consider that the two sectors are homogeneous from a classification point of view, although they may be rather different since they belong to two different regions.

Since data availability on input–output information at sector level is limited, an alternative solution is to form a similarity matrix based on technological specialization indicators (Van Stel and Nieuwenhuijsen, 2004) as suggested in Costantini *et al.* (2011) with the bilateral innovation spillovers (ts_k^{rs}) for each *k-th* sector from the *s-th* region to the *r-th* region initially un-weighted by the geographical distance included.

The resulting matrix of spillovers for each *k-th* sector is then synthesized into a linear vector by using geographical distances for aggregating the *s-th* elements. The geographical distances adopted here are calculated as the number of kilometres between the economic centres in each region bilaterally, by using the automatic algorithm based on highway distances with the shortest time criterion adopted by the Italian Automobile Association (ACI) which is the national official reference for distance calculation.[7]

Following Bode (2004), we have tested several alternative criteria for transforming geographical distances into spatial weights. Since there is no a priori information for which spatial regime should be preferred, we have considered three different plausible regimes: (a) the binary contiguity concept where only neighbouring regions matter for knowledge spillovers; (b) the *k* nearest neighbours concept (testing a bound *k* distance of 300 km); (c) the pure inverse distances.

The inclusion of innovation variables built on patent data reduced the number of NAMEA sectors in the analysis to 12, forcing us to exclude the 'Electricity, gas and water supply' sector (E in NACE codes); this is why we have calculated emissions from electricity consumption for each sector as a measure of indirect emissions (remembering that NAMEA only provides direct emissions). In this way, emissions associated with the 'Electricity, gas and water supply' sector can be easily excluded when accounting for emissions due to energy consumption indirectly at sector level.[8]

Since we are arguing that environmental performance may well be affected by agglomeration effects associated with a cluster-based choice of the adopted production technique, the term (es_k^r) related to environmental spillovers in equation (3) has been proxied by the emission intensity of the same sector into the other regions. To this end, we have built the environmental spillovers as the sum of sectoral emissions per unit of value added from the other regions (e_k^s) valid for $\forall s \neq r$, weighted by distances expressed in the same three different regimes (D_1, D_2 and D_3) mentioned above.[9] To some extent, we can interpret this variable as a sign of agglomerative effects for each sector related to the technological frontier adopted. If, *ceteris paribus*, firms are located in one region surrounded by regions where firms adopt polluting production technologies, the probability that firms will adopt cleaner production technologies will decrease so that a sort of polluting firm cluster emerges for selected geographical areas independently of the specific sector under investigation. Coherently with technological spillovers, the environmental spillovers have been tested with three different spatial regimes.

Finally, since environmental policies are considered drivers of environmental performance in equation (3), we can proxy them by the stringency of the environmental regulatory framework at regional level. In this chapter, we have proxied the regulatory efforts by using the incidence of environmental regulation on average regional income as suggested by Costantini and Crespi (2011) at a more aggregated national level. In our dataset we are not able to model specific effects for different sectors and we can only consider an overall regional environmental regulatory framework which allows a fixed structural effect to be shaped.

Environmental regulation is then represented by three alternative public expenditure measures,[10] related to current, capital and R&D expenditures for environmental protection activities as emerging from accounting documents of each region (ISTAT, 2007).[11]

Environmental performance in the Italian regions

For the sake of simplicity, in the shift-share analysis we restrict comments to main regions and five pollutants (CO_2, SO_x, NO_x, PM_{10}, NMVOC). Table 3.1 shows how Italian regions behave with respect to the national average when emission intensities are compared before they are decomposed, whereas Table 3.2 shows a quite clear North–South divide which corresponds to the well-known economic divide that historically exists between Northern and Southern Italian regions.

Nevertheless, it also shows that some Central and Southern regions (Lazio and Campania) behave quite well whereas some rich industrial regions (Veneto, Friuli Venezia Giulia) do not perform so satisfactorily, highlighting idiosyncrasies and criticalities that may be related to more complex issues that bring together geographical, economic and policy issues.

If we examine the industry mix and efficiency components, interesting insights emerge from the heterogeneous effects related to these two features.

With regard to the industry mix, Figure 3.1 clearly shows that while it is clear that more industrialized regions in the North are penalized by this structural component (Lombardia, Emilia-Romagna, Veneto, three main industrialized regions), Southern regions benefit, environmentally speaking, from their less industrialized specialization.[12]

It is also significant that one of the largest regions, Lazio (the region of Rome), as a service-oriented region benefits from its productive structure in environmental terms, and two small but economically important regions in the

Table 3.1 Regional performance: no. of pollutants out of 10 with a better performance than the national average

10 out of 10	Marche (C), Lazio (C) and Campania (C)
9 out of 10	Trentino Alto Adige (NE)
8 out of 10	Lombardia (NW) and Toscana (C)
7 out of 10	Piemonte (NW), Valle d'Aosta (NW) and Liguria (NW)
6 out of 10	Emilia-Romagna (NE) and Abruzzo (C)
5 out of 10	Veneto (NE)
4 out of 10	Calabria (S)
3 out of 10	Molise (S) and Sicilia (S)
2 out of 10	Friuli-Venezia Giulia (NE) and Umbria (C)
1 out of 10	Puglia (S) and Basilicata (S)
0 out of 10	Sardegna (S)

Note
Regional areas in brackets: NW = North West; NE = North East; C = Centre; S = South and Islands.

Table 3.2 CO_2 and SO_x emission intensity (kg × 1M€ of value added, increasing order)

Region	*CO_2*	*Region*	*SO_x*
Trentino Alto Adige	136	Trentino Alto Adige	39
Campania	141	Valle d'Aosta	45
Valle d'Aosta	153	Abruzzo	69
Piemonte	185	Campania	78
Lazio	204	Lombardia	99
Marche	206	Lazio	101
Lombardia	209	Marche	108
Abruzzo	258	Piemonte	108
Veneto	267	Calabria	123
Emilia Romagna	270	Basilicata	224
Toscana	278	Emilia-Romagna	226
ITALY	**301**	Molise	276
Calabria	307	Veneto	300
Umbria	342	**ITALY**	**315**
Friuli Venezia Giulia	353	Toscana	349
Basilicata	430	Umbria	373
Liguria	472	Friuli Venezia Giulia	539
Sicilia	547	Puglia	859
Molise	689	Liguria	886
Sardegna	824	Sicilia	1,347
Puglia	971	Sardegna	1,530

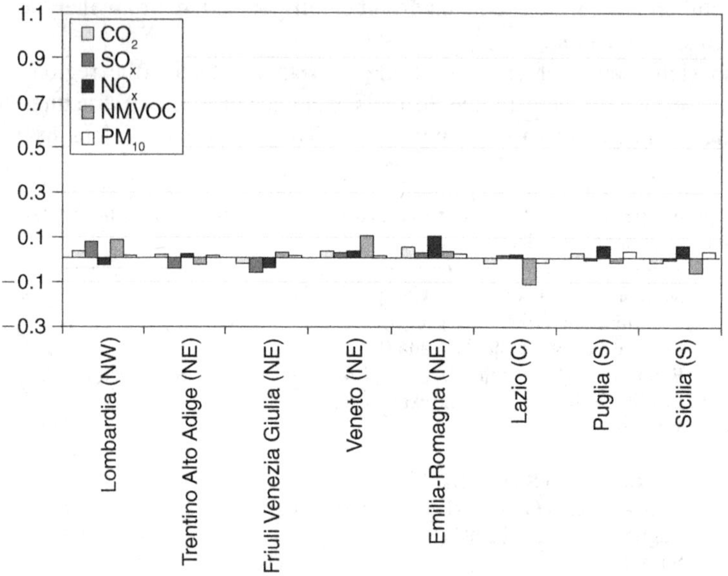

Figure 3.1 Industry mix component from shift-share analysis (coefficient *m*).

Note
NW=North West; NE=North East, C=Centre, S=South and Islands.

North, with a high degree of fiscal and legislative autonomy and cultural idiosyncrasies (including regional languages), such as Trentino Alto Adige and Friuli Venezia Giulia, also benefit on average from the industry mix component.

Summing up, this part of the shift-share analysis tells us that the North–South divide regarding industrial development obviously affects the environmental comparative advantage of a region, all other things being equal. But this is only one side of the story since this part of the analysis is not enough to disentangle clustering effects on environmental performance.

To some extent, the efficiency gap seems to be the main driving force behind regional comparative advantage showing various cases of best and worst situations that highlight how efficiency and North–South structural differences are jointly relevant to explaining different striking performances (Figure 3.2).

It is noteworthy that Friuli Venezia Giulia, a developed industrialized region associated with high income per capita, performs badly on average, and not because of its industry mix, as we commented on above, but because of specific inefficiency features. The North-East as a whole, an area of the country with high economic performance driven by export-intensive manufacturing and some heavy industry, appears to perform worse than the North-West (Piemonte and Lombardia).[13] The former is currently the region that always performs better than average with regard to both industry mix and efficiency.

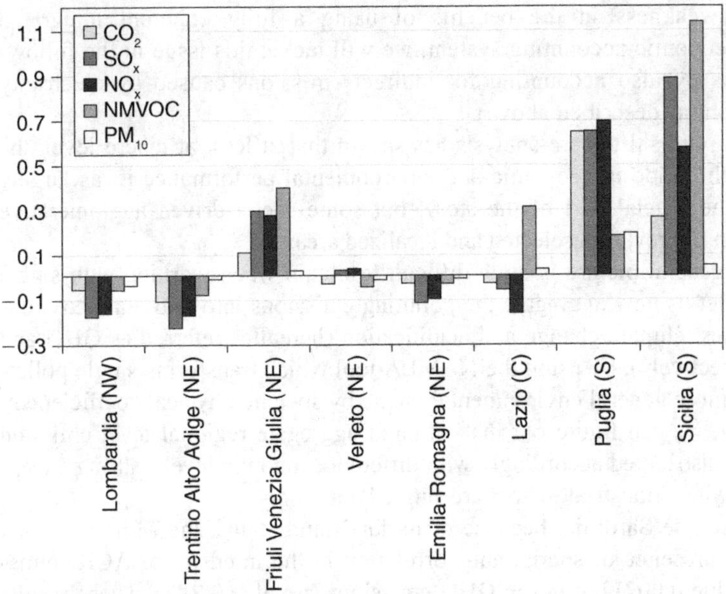

Figure 3.2 Efficiency component from shift-share analysis (coefficient *p*).

Note
NW = North West; NE = North East, C = Centre, S = South and Islands.

In other Northern industrial regions, on average, although not for all emissions, efficiency gains tend to compensate for unfavourable industry mix features. Given the often proposed dichotomy between the type of industrial development in the North-East of Italy, based on small and medium enterprises (SMEs) and districts rather than on large corporate firms with outsourcing collars, it is interesting to note that at least at macro level, the economic development model based on SMEs seems to link economic and environmental performance at general level less strictly, while inducing a more localized correlation effect between agglomeration economies and environmental and innovation spillovers.

One interesting case is Friuli Venezia Giulia, which is characterized by high innovative industrial niches but also hosts industrial sites that exploit coal quite intensively. The reasoning on regional energy structure also points to the evident good performance of a region like Trentino Alto Adige, which emerges with the best gap in three out of five emissions examined (Table 3.1). This region is less industrialized than other Northern ones and also depends enormously on renewable energy (mostly hydroelectric). Energy sector is also relevant in Southern regions, but the type of energy mix drastically affects performance. We use this result to comment on the direct nature of NAMEA emissions whereas accounting for the indirect generation of emissions would partially change the results. Though we will stick to this intrinsic NAMEA feature, a weakness in the benefits of using a fully coherent integrated emission-economic accounting system, we will tackle this issue in the following sections by also accounting for indirect emissions caused by electricity consumption (as described above).

Summing up, shift-share analysis has shown that at least at macro level, the North–South divide in economic and environmental performance is, as largely expected, the crucial part of the story, but some sector-driven agglomerative effects seem to prevail in selected and localized areas.

Since a general picture is quite difficult to obtain when working with single pollutants, let us now aggregate the polluting emissions into two main environmental issues, climate change and acidification (hereafter referred as GHG and ACID, respectively), by using the NAMEA tool which transforms single pollutants into a more general environmental theme by specific physical coefficients.[14] In this way, we can figure out that while at aggregate regional level emission intensity is distributed accordingly with different economic levels, strong exceptions arise when industrial sectors are singled out.

If we exclude Sardinia, because of its far island status, the Moran's Index reveals the presence of spatial autocorrelation in the intensity of ACID emissions (p-value 0.007) but not in GHG emissions (p-value 0.283). These results indicate that the clustering phenomenon is more relevant for local pollutants since transboundary pollution plays a greater role in more restricted geographical areas.[15]

Hence, given that the geographical distribution of polluting emissions reveals in some cases a strong spatial concentration of dirty sectors in restricted areas which may not always correspond to regions with relatively less stringent environmental regulation or lower capital and innovation intensity, a deeper investigation of such a clustering process would shed some light on the driving forces influencing the emergence of dirtier or cleaner production areas.

Driving forces of environmental performance

The econometric estimations aim to investigate the relative strength of the effects associated with labour productivity and internal and external innovation drivers as well as the role of the environmental regulatory framework. In particular, we test the influence of these factors on the geographical and sectoral distribution of environmental performance for the two aggregated damage effects due to pollutant emissions, namely GHG and ACID (Tables 3.3 and 3.4, respectively), characterized by interesting differences in the diffusion paths. To some extent, the reaction from the community will be consistent with these differences since we expect the impact of knowledge externalities to be higher for more localized polluting emissions, as represented by ACID. The collective action played by consumers and firms may be more effective because the convenience to exploit innovation externalities from neighbouring areas is potentially higher.

In fact, the inducement effect on a technology path oriented towards less-polluting production processes also comes from private initiative, and not only from public enforcement, due to a stronger and more diffused perception of damages directly associated with environmental externalities. In this sense, the probability that an innovation will also be suitable for environmental protection purposes will be higher, and the probability of a higher diffusion speed will also increase when the polluter and the agent facing environmental damages are confined to a restricted geographical dimension.

Distinguished regression models have been estimated for the two environmental themes considered here in order to understand if these expected divergences are confirmed by empirical analysis. The empirical investigation relies on OLS estimations on 12 manufacturing sectors, thus reducing potential NAMEA observations from 480 to 240. Since heteroskedasticity is a concern in this kind of analyses, we have run all our regressions with the robust standard error specification.

As a first outcome, we note that the impact of labour productivity on explaining the environmental performance is rather high in both models, and the expected negative coefficient associated with this variable can be interpreted as a positive correlation between productivity and environmental efficiency gains. This result is to be expected due to the interplay of multiple drivers along the evolution of innovation, industrial and policy paths. In line with expectations and other analyses on NAMEA data in Italy (Marin and Mazzanti, 2011), this coefficient is larger for ACID than for GHG (almost doubled), since this second environmental theme is rather more complex and influenced by a broader mix of driving factors whereas the first one is more circumscribed both in geographical and sectoral terms.

Since we have disentangled pure innovation effects from all other characteristics in the production function, we can affirm that labour productivity explains all the structural features in the production process such as the adoption of environmental management systems, quality control and highly efficient mechanical appraisals which are not specifically caught by the innovative capacity of the economic sector captured by patent intensity.[16] Moreover, we have also included a specific variable related to energy intensity for each sector and we have introduced a dummy variable which absorbs the effect of specific dirty industries. In this way, productivity gains and innovation effects can be interpreted as the real impact on environmental efficiency related to investments in technology and labour productivity. Consistently with differences in the two environmental themes, sector-specific features seem to be prominent in the explanation of environmental efficiency behaviour for ACID emissions, as the coefficient capturing the effect related to the fact that one sector belongs to those classified as relatively more polluting explains a great portion of environmental performance, again more than double GHG emissions.

Second, with regard to the role of environmental efficiency spillovers, it is worth noting that they play a significant role in explaining environmental

Table 3.3 Drivers of regional environmental performance for GHG emissions

Dep var GHG	*(1)*	*(2)*	*(3)*	*(4)*	*(5)*	*(6)*	*(7)*
Labour productivity	–0.756***	–0.671***	–0.688***	–0.714***	–0.501***	–0.542***	–0.522***
	(–4.13)	(–3.85)	(–4.05)	(–4.12)	(–2.94)	(–3.17)	(–3.09)
Internal innovation	–0.009	–0.001	0.002	0.005	0.009	0.003	0.014
	(–0.33)	(–0.04)	(0.01)	(0.16)	(0.32)	(0.11)	(0.50)
Energy intensity	0.645***	0.541***	0.531***	0.549***	0.567***	0.557***	0.583***
	(14.67)	(11.64)	(12.23)	(10.63)	(11.41)	(12.31)	(10.18)
Dirty sector dummy	1.331***	0.996***	0.925***	1.033***	0.976***	0.894***	0.997***
	(12.81)	(7.33)	(6.64)	(7.17)	(7.08)	(6.31)	(6.67)
Environ. spillovers D_1		0.243***			0.236***		
		(3.84)			(3.57)		
Environ. spillovers D_2			0.289***			0.288***	
			(4.40)			(4.40)	
Environ. spillovers D_3				0.229***			0.216***
				(3.05)			(2.74)
Tech. reg. spillovers D_1					–0.125***		
					(–2.97)		
Tech. reg. spillovers D_2						–0.097**	
						(–2.57)	

Tech. reg. spillovers D_3							–0.152***
							(–2.98)
Constant	4.121***	4.083***	2.77***	4.014***	3.013***	2.184***	3.01***
	(6.77)	(6.72)	(5.01)	(6.80)	(4.67)	(3.69)	(4.91)
Regional dummies	Yes	Yes	Yes	Yes	Yes	Yes	Yes
No. obs.	209	209	209	209	209	209	209
Adj. R-sq.	0.78	0.80	0.81	0.79	0.81	0.81	0.81
F-stat	32.22	42.3	44.92	40.35	39.6	45.31	41.55
Root MSE	0.63	0.61	0.60	0.62	0.60	0.59	0.60
Hausman Chi-sq.					0.23 (0.63)	0.02 (0.89)	0.05 (0.82)
Average VIF value					1.54	1.45	1.73
LM (lag)	0.03 (0.86)	0.01 (0.94)	0.01 (0.97)	0.01 (0.97)	0.12 (0.72)	0.02 (0.89)	0.15 (0.69)
LM (error)	3.88 (0.05)	3.19 (0.07)	3.40 (0.07)	2.50 (0.11)	3.31 (0.07)	3.34 (0.07)	3.18 (0.07)
Robust LM (error)	4.95 (0.03)	3.64 (0.06)	3.94 (0.05)	2.90 (0.09)	3.33 (0.07)	3.67 (0.06)	3.12 (0.08)

Notes

***, **, *, for *p-values* of 0.01, 0.05, 0.1, respectively; robust *t-stat* values in parentheses.

Table 3.4 Drivers of regional environmental performance for ACID emissions

Dep var ACID	*(1)*	*(2)*	*(3)*	*(4)*	*(5)*	*(6)*	*(7)*
Labour productivity	−1.543***	−1.383***	−1.301***	−1.313***	−1.201***	−1.139***	−1.051***
	(−6.16)	(−5.32)	(−5.73)	(−4.76)	(−4.61)	(−4.94)	(−3.93)
Internal innovation	−0.019	−0.017	−0.013	−0.010	−0.006	−0.010	0.004
	(−0.53)	(−0.47)	(−0.36)	(−0.28)	(−0.17)	(−0.29)	(0.10)
Energy intensity	0.404***	0.373***	0.358***	0.352***	0.398***	0.389***	0.392***
	(8.97)	(7.60)	(8.01)	(6.88)	(7.59)	(8.15)	(7.18)
Dirty sector dummy	2.559***	2.272***	2.034***	2.155***	2.247***	2.008***	2.084***
	(20.76)	(9.05)	(7.03)	(8.55)	(9.03)	(6.97)	(8.46)
Environ. spillovers D_1		0.109			0.106		
		(1.35)			(1.31)		
Environ. spillovers D_2			0.195**			0.191**	
			(2.16)			(2.12)	
Environ. spillovers D_3				0.163*			0.162**
				(1.90)			(1.96)
Tech. reg. spillovers D_1					−0.134**		
					(−2.40)		
Tech. reg. spillovers D_2						−0.111**	
						(−2.29)	

Tech. reg. spillovers D_3							–0.204***
							(–3.12)
Constant	4.596***	4.423***	3.489***	4.228***	3.281***	2.833***	2.865***
	(5.41)	(5.21)	(4.47)	(4.89)	(3.51)	(3.46)	(3.28)
Regional dummies	Yes	Yes	Yes	Yes	Yes	Yes	Yes
No. obs.	209	209	209	209	209	209	209
Adj. R-sq.	0.77	0.77	0.77	0.77	0.78	0.78	0.79
F-stat	47.96	49.87	54.32	49.8	48.30	54.90	50.78
Root MSE	0.77	0.77	0.77	0.77	0.76	0.76	0.75
Hausman Chi-sq.					0.04	0.01	0.05
					(0.84)	(0.93)	(0.83)
Average VIF value					1.70	1.83	1.98
LM (lag)	0.03 (0.86)	0.02 (0.89)	0.01 (0.92)	0.01 (0.93)	0.10 (0.76)	0.07 (0.79)	0.21 (0.65)
LM (error)	0.68 (0.41)	0.71 (0.40)	0.79 (0.37)	0.44 (0.50)	0.87 (0.35)	1.12 (0.29)	1.11 (0.29)

Notes

***, **, *, for *p-values* of 0.01, 0.05, 0.1, respectively; robust *t-stat* values in parentheses.

performance especially for GHG emissions. The spatial regime where the environmental spillovers seem to have the greatest effect coincides with regions in the range of 300 km since estimated coefficients are higher for both GHG and ACID. Nonetheless, some differences emerge between the two environmental themes since for GHG, all the three spatial regimes are statistically robust and coefficient values present a small discrepancy, whereas for ACID, the D_2 spatial regime seems to be the more robust and significant.

The expected positive coefficient can be interpreted as first evidence of the existence of clusters not only intended as an agglomeration of specific sectors into restricted areas, but also as a first influence of the technology adopted in the production processes. This means that together with the agglomeration of specific sectors into restricted areas, there is also some convergence in production processes and techniques. To some extent, we can affirm that the clustering process of specific polluting sectors in relation to contiguous geographical areas may be followed by common choices in the adoption of cleaner or dirtier technologies.

On the other hand, it is worth noting that the level of internal innovation, expressed as the number of patents per value added, plays no role in explaining environmental efficiency since the coefficient presents low and no statistical robustness in all specifications. This is plausible given that our innovation variable relates to the general efforts to produce technology, without specific environmental purposes. Further research steps could be to consider specific environmental innovation rather than a general innovative capacity, when the efforts by the OECD and WIPO will be conducive to a well established and consolidated methodology to classify patents for environmental protection purposes (OECD, 2008).

On the contrary, technological interregional spillovers seem to play a more effective role in improving environmental efficiency. The higher impact of innovation spillovers compared with internal innovation can be explained by the nature of our innovation variable which refers to a general innovation output. In this case, we can affirm that the higher the knowledge flows from other regions, the more likely the availability of environment-friendly technologies and the higher the reduction in emission intensity. The portfolio of innovations available within a sector at national level, similar to the business group effect for firms as emphasized by Belenzon and Berkowitz (2007), could extend the set of innovation choices at regional level. Firms belonging to a defined sector can eventually find the environment-friendly innovations they need at national level.

In the case of innovation spillovers, the three spatial regimes all give robust results, meaning that innovation effects spread out of the regional borders with no limit distance. Nonetheless, the highest effect is associated to the D_3 regime, meaning that the higher the availability of technological innovation at sector level, the more likely the capacity of each sector to choose the best environment-friendly technology and the better the environmental performance.

Consistently with our expectations, the positive influence of technological spillovers on environmental performance is higher for more localized pollutants (ACID) since the collective reaction to better perceived environmental damage will be to adopt the innovations available in each sector more rapidly and diffusely. In this case, the size of the coefficient – its economic significance – is larger compared with GHG, also confirming evidence previously found for labour productivity.[17]

Since spatial correlation may give biased results, we have implemented specific checks by implementing diagnostics for lag and error spatial dependence.[18] As Lagrange Multipliers (LM) tests for the existence of both spatial lag or spatial error reveal, only a weak spatial correlation emerges from the LM error test for GHG estimation whereas for ACID, OLS results are all robust.[19]

Finally, with regard to the role of environmental regulation, we tested the role of the three alternative measures considered here (current, capital and R&D public expenditures for environmental protection at regional level) with one temporal lag, taking all spatial regimes for GHG and ACID alternatively (Table 3.5). The choice for the temporal dimension is quite obvious since the regulatory framework may induce firms to be more environmentally responsible only after a certain period required to adapt production processes to new rules has passed.

Although previous findings do not change when the regulatory effort is included, some interesting differences emerge when comparing the two environmental themes. All coefficients show an expected negative sign since an increase in the social price of negative externalities would force firms to adopt more efficient production processes, but for GHG only R&D public expenditures for environmental protection seems to positively influence environmental performance.

On the contrary, the regulatory framework seems to be more effective for ACID emissions since all the three measures have a positive influence on environmental efficiency gains with robust statistical significance. Here, too, empirical results seem to be in line with expectations since the capacity of the collective action to force the local government to adopt more stringent environmental standards and rules is more effective when there is a higher perception of damage from the community.

Evidence on the GHG emissions side can also be easily explained if we think about the well-known weakness of Italian environmental policy which does not present a structural, clear and long-term strategy for addressing climate change.

As a final robustness check, we also tested the potential effects of neighbouring environmental regulatory system in line with Gray and Shadbegian (2007), but we did not find any significant effect on emission intensity reduction.

Conclusions

The achievement of positive environmental performance at national level could strongly depend on differences in local capabilities of both institutions and the

Table 3.5 The role of environmental regulation

	GHG			*ACID*		
	(1)	*(2)*	*(3)*	*(4)*	*(5)*	*(6)*
Labour productivity	–0.501***	–0.542***	–0.522***	–1.201***	–1.139***	–1.051***
	(–2.94)	(–3.17)	(–3.09)	(–4.61)	(–4.94)	(–3.93)
Internal innovation	0.009	0.003	0.014	–0.006	–0.01	0.004
	(0.32)	(0.11)	(0.50)	–(0.17)	(–0.29)	(0.10)
Energy intensity	0.567***	0.557***	0.583***	0.398***	0.389***	0.392***
	(11.41)	(12.31)	(10.18)	(7.59)	(8.15)	(7.18)
Dirty sector dummy	0.976***	0.894***	0.997***	2.247***	2.008***	2.084***
	(7.08)	(6.31)	(6.67)	(9.03)	(6.97)	(8.46)
Environ. spillovers D1	0.236***			0.106		
	(3.57)			(1.31)		
Environ. spillovers D2		0.288***			0.191**	
		(4.40)			(2.12)	
Environ. spillovers D3			0.216***			0.162**
			(2.74)			(1.96)
Tech. reg. spillovers D1	–0.125***			–0.134**		
	(–2.97)			(–2.40)		

Tech. reg. spillovers D2		–0.097**			–0.111**	
		(–2.57)			(–2.29)	
Tech. reg. spillovers D3			–0.152***			–0.204***
			(–2.98)			(–3.12)
Env. reg. current exp.	–0.105			–0.62**		
	(–0.81)			(–2.05)		
Env. reg. capital exp.		–0.005			–0.272**	
		(–0.03)			(–2.03)	
Env. reg. R&D exp.			–0.163**			–0.288**
			(–2.58)			(–2.27)
Constant	2.95***	2.187***	2.527***	4.738***	4.143***	1.84**
	(4.54)	(3.54)	(3.85)	(5.78)	(5.90)	(2.11)
Regional dummies	Yes	Yes	Yes	Yes	Yes	Yes
No. obs.	209	209	209	209	209	209
Adj. R-sq.	0.81	0.81	0.80	0.77	0.78	0.79
F-stat	39.60	45.31	41.55	48.30	54.90	50.78
Root MSE	0.60	0.59	0.60	0.76	0.76	0.75

private business sector. The decomposition of economic-environment accounting in industry specialization and efficiency components revealed by shift-share analysis tells us that the Italian North–South divide regarding industrial development and productive specialization patterns obviously affects regional environmental performance. On the one hand, such strong North–South differences in performance may reflect coherence with economic development stages and priorities but, on the other hand, can also signal regulatory and industrial policy failures or successes occurring in different regions even at similar income levels. Industrial regional specialization matters but efficiency effects also play a crucial role. The North-East as a whole, a leading economic area in the country with high economic performance driven by export-intensive manufacturing sectors, appears to perform worse than the Western part of the industrialized North. Traditional elements of the North–South divide are not therefore an exhaustive explanation of the heterogeneous geographical distribution of pollution in Italy.

Empirical investigation into which drivers may explain such distribution through econometric analysis reveals that innovation and environmental efficiency spillovers are highly relevant. For a more global environmental theme such as GHG emissions in particular, it is worth noting that environmental spillovers play a significant role in explaining environmental performance. This result can be interpreted as first evidence of the existence of clusters not only intended as an agglomeration of specific sectors into restricted areas, but also as the existence of a common technology adopted in production processes. The clustering process of specific polluting sectors into selected geographical areas seems to be followed by common choices in the adoption of cleaner or dirtier technologies, which helps us to explain why the same sector specialization into different regions may be characterized by different emission intensities or efficiency as found in the shift-share analysis.

A second important result is that technological interregional spillovers seem to play a more effective role in improving environmental efficiency than internal innovation, with an increasing effect for more localized pollutants. The greater overlapping between polluters and agents perceiving environmental damage for more localized emissions also explains the stronger effectiveness of environmental regulation at regional level in fostering environmental efficiency gains.

Appendix

Table 3.A.1 Productive branches (ATECO 2001) and NACE code

Productive branches	*NACE code*
Agriculture, hunting and forestry	A
Fishing	B
Mining and quarrying	C
Manufacture of food products, beverages and tobacco	DA
Manufacture of textiles and textile products	DB
Manufacture of leather and leather products	DC
Manufacture of wood and wood products, Manufacture of rubber and plastic products, Manufacturing n.e.c.	DD-DH-DN
Manufacture of pulp, paper and paper products	DE
Manufacture of coke, refined petroleum products and nuclear fuel, Manufacture of chemicals, chemical products and man-made fibres	DF-DG
Manufacture of other non-metallic mineral products	DI
Manufacture of basic metals and fabricated metal	DJ
Manufacture of machinery and equipment n.e.c., Manufacture of electrical and optical equipment, Manufacture of transport equipment	DK-DL-DM
Electricity, gas and water supply	E
Construction	F
Wholesale and retail trade; repair of motor vehicles, motorcycles and personal and household goods	G
Hotels and restaurants	H
Transport, storage and communication	I
Financial intermediation	J
Real estate, renting and business activities	K
Public administration and defense; compulsory social security	L
Education	M
Health and social work	N
Other community, social and personal service activities	O
Household related activities	P
Total	

Table 3.A.2 Concordance classification for NACE sectors, NAMEA sectors and IPC codes

CODE NAMEA	*CODE NACE*	*CODE IPC*
1	A – Agriculture	A01
3	C – Mining and quarrying	E21
4	DA15 – Manufacture of food products and beverages DA16 – Manufacture of tobacco products	A21-A22-A23-A24-C12-C13
5	DB17 – Manufacture of textiles DB18 – Manufacture of wearing apparel; dressing; dyeing of fur	A41-A42-D01-D02-D03-D04-D05-D06
6	DC19 – Tanning, dressing of leather; Manufacture of luggage	A43-B68-C14
7	DD20 – Manufacture of wood and of products of wood and cork, except furniture; Manufacture of articles of straw and plaiting materials DH25 – Manufacture of rubber and plastic products DN36 – Manufacture of furniture; Manufacturing n.e.c.	A44-A45-A46-A47-A63-B09-B27-B29-C02-C30-G10
8	DE21 – Manufacture of pulp, paper and paper products DE22 – Publishing, printing, reproduction of recorded media	B31-B42-B43-B44-D21-G09
9	DF23 – Manufacture of coke, refined petroleum products and nuclear fuel DG24 – Manufacture of chemicals and chemical products	C01-C05-C06-C07-C08-C09-C10-C11-C40-F16

10	DI26 – Manufacture of other non-metallic mineral products	B28-B32-C03-C04
11	DJ27 – Manufacture of basic metals DJ28 – Manufacture of fabricated metal products, except machinery and equipment	B25-B26-C21-C22-C23-C25-D07-E02-E05
12	DK29 – Manufacture of machinery and equipment n.e.c. DL30 – Manufacture of office machinery and computers DL31 – Manufacture of electrical machinery and apparatus n.e.c. DL32 – Manufacture of radio, television and communication equipment and apparatus DL33 – Manufacture of medical, precision and optical instruments, watches and clocks DM34 – Manufacture of motor vehicles, trailers and semi-trailers DM35 – Manufacture of other transport equipment	A61-A62-B01-B02-B03-B04-B05-B06-B07-B08-B21-B22-B23-B24-B30-B41-B60-B61-B62-B63-B64-B65-B66-B67-B81-B82-F01-F02-F03-F04-F15-F21-F23-F24-F25-F26-F27-F41-F42-G01-G02-G03-G04-G05-G06-G07-G08-G11-G12-H01-H02-H03-H04-H05
13	E – Electricity, gas and water supply	E03-F17-F22-F28-G21-H02
14	F – Construction	E01-E04-E06

Source: own elaborations on Schmoch *et al.* (2003).

Table 3.A.3 Variables description

Variable	Description
Labour productivity	Value added per full-time equivalent job unit
Environ. spillovers (D1)	Sector-specific pollutant emissions in directly neighbouring regions eq. [14]
Environ. spillovers (D2)	Sector-specific pollutant emissions in regions ≤300 km maximum distance eq. [15]
Environ. spillovers (D3)	Sector-specific pollutant emissions in all regions eq. [16]
Energy intensity	Electricity consumption to value added ratio for each specific sector
Env. reg. current exp.	Environmental regional expenditure 2004 to value added ratio (current)
Env. reg. capital exp.	Environmental regional expenditure 2004 to value added ratio (capital)
Env. reg. R&D exp.	Environmental R&D regional expenditure to value added ratio 2004
Internal innovation	Number of patents per value added; five-year average 2000–2004
Tech. reg. spillovers (D1)	Sector-specific innovation spillovers from patents intensity (five-year average 2000–2004) available in directly neighbouring regions eq. [11]
Tech. reg. spillovers (D2)	Sector-specific innovation spillovers from patents intensity (five-year average 2000–2004) available in regions ≤300 km maximum distance eq. [12]
Tech. reg. spillovers (D3)	Sector-specific innovation spillovers from patents intensity (five-year average 2000–2004) available in all regions eq. [13]
Dirty sector dummy	Dummy for heavy polluting sectors as explained in note 10

Notes

1 The first NAMEA was developed by the Dutch Central Bureau of Statistics (De Boo *et al*., 1993), and earlier contributions such as Ike (1999), Keuning *et al*. (1999), Steenge (1999) and Vaze (1999) provided empirical analyses related to the possible policy implications deriving from sector-specific environmental performance. In the NAMEA tables, environmental pressures, in particular air emissions, and economic data (value added, final consumption expenditures and full-time equivalent employment) are assigned to the economic branches of resident units directly responsible for environmental and economic phenomena.

2 The technological effect argues that economic sectors may adopt less-polluting technologies, either because of market-driven technological progress or government regulation, as emphasized by Cole *et al*. (2005).

3 A well consolidated literature recognizes that productivity is the core economic driver explaining environmental performance, relying on the so-called environmental Kuznets curves (EKC) realm where an inverted U-shaped curve may theoretically represent linkages between economic development and environmental performance (Andreoni and Levinson, 2001). For an extensive literature review on EKC, see, for instance, Costantini and Martini (2010) and Dinda (2004).

4 A detailed description of the methodology used can be found in Costantini *et al*. (2011).

5 See Table 3.A.1 in the Appendix for the list of sectors and NACE codes considered.

6 To calculate the total GHG emissions, the CO_2, CH_4 and N_2O emissions are converted into tonnes of CO_2 equivalent, by multiplying each gaseous emission for the corresponding Global Warming Potential (GWP). To aggregate the different pollutant emissions (NO_x, SO_x and NH_3) that contribute to the acidifying phenomenon, the specific Potential Acid Equivalent (PAE) corresponding to each one is considered.

7 The official distances provided by ACI are computed in order to give a homogeneous criterion for funding business travel costs, thus representing the best available proxy for costs of face-to-face contacts which are recognized as the main channel for regional knowledge spillovers.

8 For details on calculation methodology for indirect emissions from electricity consumption, see Costantini *et al*. (2011).

9 The distance regime 1 (D_1) is related to neighbouring regions (first-order binary contiguity); the second distance regime (D_2) reflects a maximum distances between regions equal to 300 km (k-nearest neighbours); the last distance regime is linked to the inverse distances between regions. A detailed description of the three spatial regimes can be found in Costantini *et al*. (2011).

10 See Table 3.A.3 for details regarding all the considered variables. Measures are officially provided by the environmental accounting unit of ISTAT.

11 We acknowledge the existence of aggregation issues because our regulation measures should ideally be at sector level, but no data are available at the moment for Italian regions with sector specification.

12 All detailed results of the shift-share analysis are available upon request from the authors.

13 The most industrialized Italian regions are definitely Lombardia, Veneto and Emilia-Romagna, with an industrial GDP share of around 33–34 per cent, whereas Piemonte and Friuli Venezia Giulia are relatively less industrialized.

14 For details on specific converting coefficients for all pollutants, see the technical notes on NAMEA available from De Boo *et al*. (1993).

15 The (univariate) Moran's I measures the type and strength of spatial autocorrelation from spatial interaction effects (e.g. externalities or spillover effects) in a data distribution. This statistic determines the extent of linear association between the values in a given location with values of the same variable in neighbouring locations.

16 The specific dirty industries assuming value 1 in the dummy are: Agriculture, Manufacture of coke, refined petroleum products and nuclear fuel, Manufacture of chemicals and chemical products, Manufacture of other non-metallic mineral products.

17 In order to check for the robustness of our model, we tested both potential multicollinearity of internal and external innovation as well as potential endogeneity of the covariate explaining regional innovation by performing the Variance Inflation Factor (VIF) and the Hausman test. Average VIF values and Hausman statistics confirm the robustness of our specification.

18 The spatial weights matrix used to test the presence of spatial dependence is based on a *rook* weights matrix (a contiguity-based matrix) for the Italian regions initially calculated with the Geoda 0.9.5-i software. For Italian regions, the *queen* weights matrix (that considers borders and vertices) is equal to the *rook* one (that considers only borders). However, further work has been done because our dataset not only consists of 20 statistical geo-units (regions) but 209 statistical units (19 regions – because Sardinia has been excluded considering its far island status – times 11 NAMEA sectors). As suggested by Anselin (personal correspondence), a 'trick' to obtain such a spatial weights matrix is to replicate the initial one – opportunely recoded each time – for the number of considered sectors. Thus the final weights matrix has the same number of observations as the considered cross-sector–region dataset.

19 For GHGs we checked if a spatially corrected model produces different results, but all coefficients remain unchanged in statistical significance.

References

Acs, Z.J., Anselin, L. and Varga, A. (2002) 'Patents and innovation counts as measures of regional production of new knowledge', *Research Policy*, 31: 1069–1085.

Andreoni, J. and Levinson, A. (2001) 'The simple analytics of the environmental Kuznets curve', *Journal of Public Economics*, 80: 269–286.

Antonelli, C., Patrucco, P.P. and Quatraro, F. (in press) 'Pecuniary knowledge externalities: evidence from European regions', *Economic Geography*.

Belenzon, S. and Berkowitz, T. (2007) 'Innovation in business groups', CEP discussion paper 833, LSE, London.

Bode, E. (2004) 'The spatial pattern of localized R&D spillovers: an empirical investigation for Germany', *Journal of Economic Geography*, 4: 43–64.

Boschma, R. and Frenken, K. (2009) 'Some notes on institutions in evolutionary economic geography', *Economic Geography*, 85: 151–158.

Boschma, R. and Iammarino, S. (2009) 'Related variety, trade linkages, and regional growth in Italy', *Economic Geography*, 85: 289–311.

Cefis, E., Rosenkranz, S. and Weitzel, U. (2009) 'Effects of coordinated strategies on product and process R&D', *Journal of Economics*, 96: 193–222.

Cole, M.A., Elliott, R. and Shimamoto, K. (2005) 'Industrial characteristics, environmental regulations and air pollution: an analysis of the UK manufacturing sector', *Journal of Environmental Economics and Management*, 50: 121–143.

Costa, M. and Iezzi, S. (2004) 'Technology spillover and regional convergence process: a statistical analysis of the Italian case', *Statistical Methods & Applications*, 13: 375–398.

Costantini, V. and Crespi, F. (in press) 'Public policies for a sustainable energy sector: regulation, diversity and fostering of innovation', *Journal of Evolutionary Economics*.

Costantini, V. and Martini, C. (2010) 'A modified environmental Kuznets curve for sustainable development assessment using panel data', *International Journal of Global Environmental Issues*, 10: 84–122.

Costantini, V., Mazzanti, M. and Montini, A. (2011) 'Environmental performance, innovation and regional spillovers', Working Papers 2011/03, University of Ferrara, Department of Economics Institutions and Territory (DEIT).

De Boo, A., Bosch, P., Gorter, C.N. and Keuning, S.J. (1993) 'An environmental module and the complete system of national accounts', in A. Franz and C. Stahmer (eds), *Approaches to Environmental Accounting*, Heidelberg: Physica Verlag.

de Haan, M. (2004) 'Accounting for goods and bads: measuring environmental pressure in a national accounts framework', mimeo, Voorburg: Statistics Netherlands.

de Haan, M. and Keuning, S.J. (1996) 'Taking the environment into account: the NAMEA approach', *Review of Income and Wealth*, 42: 131–148.

Dinda, S. (2004) 'Environmental Kuznets curve hypothesis: a survey', *Ecological Economics*, 49: 431–455.

Esteban, J. (1972) 'A reinterpretation of shift-share analysis', *Regional Science and Urban Economics*, 2: 249–261.

Esteban, J. (2000) 'Regional convergence in Europe and the industry mix: a shift-share analysis', *Regional Science and Urban Economics*, 30: 353–364.

Femia, A. and Panfili, P. (2005) 'Analytical applications of the NAMEA', paper presented at the annual meeting of the Italian Statistics Society, Rome.

Frenken, K., Van Oort, F. and Verburg, T. (2007) 'Related variety, unrelated variety and regional economic growth', *Regional Studies*, 41: 685–697.

Gibbs, D. (2006) 'Prospects for an environmental economic geography: linking ecological modernization and regulationist approaches', *Economic Geography*, 82: 193–215.

Gray, W.B. and Shadbegian, R.J. (2007) 'The environmental performance of polluting plants: a spatial analysis', *Journal of Regional Science*, 47: 63–84.

Ike, T. (1999) 'A Japanese NAMEA', *Structural Change and Economic Dynamics*, 10: 123–149.

ISTAT (2007) *La ricerca e sviluppo in Italia: Anno 2004*. Online: www.istat.it/dati/dataset/20070329_00.

ISTAT (2009) *Namea: emissioni atmosferiche regionali*, Contabilità Ambientale, Rome: ISTAT.

Keuning, S., van Dalen, J. and de Haan, M. (1999) 'The Netherlands' NAMEA: presentation, usage and future extensions', *Structural Change and Economic Dynamics*, 10: 15–37.

Los, B. (2000) 'The empirical performance of a new inter-industry technology spillover measure', in P.P. Saviotti and B. Nooteboom (eds), *Technology and Knowledge*, Cheltenham: Edward Elgar, pp. 118–151.

Marin, G. and Mazzanti, M. (in press) 'The evolution of environmental and labour productivities dynamics', *Journal of Evolutionary Economics*.

Mazzanti, M. and Montini, A. (2010a) 'Embedding emission efficiency at regional level: analyses of NAMEA data', *Ecological Economics*, 69: 2457–2467.

Mazzanti, M. and Montini, A. (2010b) *Environmental Efficiency, Innovation and Economic Performances*, Abingdon: Routledge.

Mazzanti, M., Montini, A. and Zoboli, R. (2008) 'Economic dynamics, emission trends and the EKC hypothesis: new evidence using NAMEA data for Italy', *Economic Systems Research*, 20: 279–305.

OECD (2008) *Environmental Policy, Technological Innovation and Patents*, OECD Studies on Environmental Innovation, Paris: OECD.

Schmoch, U., Laville, F., Patel, P. and Frietsch, R. (2003) *Linking Technology Areas to Industrial Sectors*, Final Report to the European Commission, DG Research, Brussels: European Commission.

Steenge, A. (1999) 'Input–output theory and institutional aspects of environmental policy', *Structural Change and Economic Dynamics*, 10: 161–176.

Van Stel, A.J. and Nieuwenhuijsen, H.R. (2004) 'Knowledge spillovers and economic growth: an analysis using data of Dutch regions in the period 1987–1995', *Regional Studies*, 38: 393–407.

Vaze, P. (1999) 'A NAMEA for the UK', *Structural Change and Economic Dynamics*, 10: 99–121.

4 Development and use of a regional NAMEA in Emilia-Romagna (Italy)

Elisa Bonazzi and Michele Sansoni

Introduction

The development and use of environmental accounting tools in Emilia-Romagna Italy starts by considering the need to integrate conventional economic indicators when drawing up sustainability reports and monitoring the effects of regional policies. So far, this study has been focused mainly on RAMEA (a regional NAMEA), its updating and related developments. Research activities have focused on two main fields: the extension of the framework to new environmental issues and the possibility of updating data and the use of RAMEA as a tool for regional environmental reports and environmental assessments of regional plans.

This chapter attempts to show the structure of the RAMEA matrix as developed in Emilia-Romagna and introduce two further developments: the results of an application of a statistical analysis (shift-share) and the extensions made with environmental taxes, industrial waste production and energy consumptions. The strategic aim is to support policy-making, also at regional level, providing an evidence base for sustainable policies and thus integrating sustainable development concerns at all levels.

There has been increasing worldwide interest in developing a broader set of statistics that gives values to aspects left outside the traditional economic system. Countries and governments need to develop a more comprehensive view of progress rather than focusing mainly on essential economic indicators such as gross domestic product (GDP). Non-market factors like environmental externalities are not counted in the GDP and conventional economic indicators. From international to local scales, there is a growing emphasis on 'evidence-based policy-making' which needs better measures of the current outcomes of programmes and policies, thus requiring statistical and analytical approaches that go beyond national borders and the conventional reporting system.[1] The Revised European Strategy for Environmental Accounting (ESEA; Eurostat, 2008) will help to ensure the availability of important environmental accounts data from all European countries and will enable these data to be harmonized, timely and of adequate quality in order to facilitate their use in developing and informing policy. In addition, the ESEA Task Force recommends that the priority for environmental accounts focuses primarily on physical and monetary flows including hybrid accounts such as NAMEA,[2] economic information on the environment and economic activities and products related to the environment and other environmentally related transactions such as taxes and subsidies.

State of the art of RAMEA matrix in Emilia-Romagna

If we consider GDP Emilia-Romagna is one of the richest regions in Italy. Nevertheless in Emilia-Romagna, as in many other developed regions, there is a critical growth of pollutant emissions (in particular greenhouse gases – GHG); transport, industries, agriculture and residential consumptions are the main drivers of this growth. To tackle these problems, a low-carbon and green economy is included in the new Regional Development Strategy. The research activities of compilation and development of a Regional NAMEA (RAMEA) matrix in Emilia-Romagna starts with the European project 'RAMEA – Regionalized NAMEA-type matrix' financed by the INTERREG IIIC Program 2005–2007 under the Regional Framework Operation 'GROW'. The project, promoted by the Emilia-Romagna Region and led by Arpa Emilia-Romagna (the Regional Environment Agency), involved seven partners from four European regions, who cooperated for two years to build four regional NAMEAs: Emilia-Romagna in Italy, South-East England, Noord Brabant in the Netherlands and Malopolska in Poland (Bonazzi *et al.*, 2008; Sansoni *et al.*, 2010). The main outcomes of this project in Emilia-Romagna[3] are the following: two RAMEA air emission account matrices extended with input–output tables (for 1995 and 2000); a guideline book describing the compiling process and regional case studies (RAMEA, 2007); a shift-share analysis built on the year 2000 matrix (Dosi *et al.*, 2008). In detail, the RAMEA project can be considered the first example of four EU regions that work together to build a regional NAMEA by following a shared methodology and improving a knowledge base for regional sustainable development policies: the regional scale for economic-environmental accounting seems to demonstrate a crucial role in building a pathway for sustainable development.

Based on these premises, the development of RAMEA matrices is encouraged by the European Commission Environment DG (European Commission, 2008) and now funded in Emilia-Romagna by the regional government. The research activity of Arpa Emilia-Romagna focuses on two main fields: the extension of the framework to new environmental issues and the possibility of updating data and the use of RAMEA as a tool for Regional Environmental Reports and Environmental Assessments.

Linking environmental and economic indicators could encourage and facilitate the involvement of the decision-makers who are likely to be more familiar with socio-economic concepts, but who are going to pay an increasing amount of attention to the effects of economic activities on the environment.

Since application to policies is a fundamental requisite for environmental accounting tools that aspire to be more than just a mere compilation of data,

RAMEA has been conceived as 'a multi-purpose information system which is able to inform the public and policy-makers about the status quo of the environmental assets and environmental pollution', useful for organizing and analysing economic and environmental data in relation to policy objectives (Goralczyk and Stauvermann, 2007).

RAMEA is based on an internationally accepted methodology (UN, Eurostat), reliable data (official statistical accounts) and standardized systems (SEEA, SNA and ESA). These conditions ensure its consistency with similar tools at national level (NAMEA). The economic activities follow NACE classification and the Household categories COICOP nomenclature. The RAMEA framework proposed for Emilia-Romagna is shown in Figure 4.1.

RAMEA could be scheduled for different kinds of analyses to explore some of the possibilities that this type of tool offers to regional planning/reporting (e.g. monitoring regional economic performance, air emissions and eco-efficiency, comparing regional eco-efficiency with national eco-efficiency and understanding the indirect effects/responsibilities of production and consumption chains on the environment).

Shift-share analysis

If we consider RAMEA for the year 2000, the intensity of emission of GHG in the regional economic system, compared with the national average, has been analysed. The indicator 'intensity of emission', explained as the ratio between GHG emissions and value added, has been used as a measure of eco-efficiency.[4]

By means of shift-share analysis, the role of the productive structure as a cause in the average gap between Emilia-Romagna and Italian efficiency of emissions has been isolated and quantified and a measure of the role of the specific efficiency of emissions of productive fields obtained in a complementary way (Bonazzi *et al.*, 2008; Bonazzi and Sansoni, 2008; Dosi *et al.*, 2008; Sansoni *et al.*, 2010). The approach on how to derive and analyse the shift-share signs, focusing on pollution-related issues, follows Mazzanti *et al.* (2007) and Mazzanti and Montini (2009). The choice of this methodology derives from a search

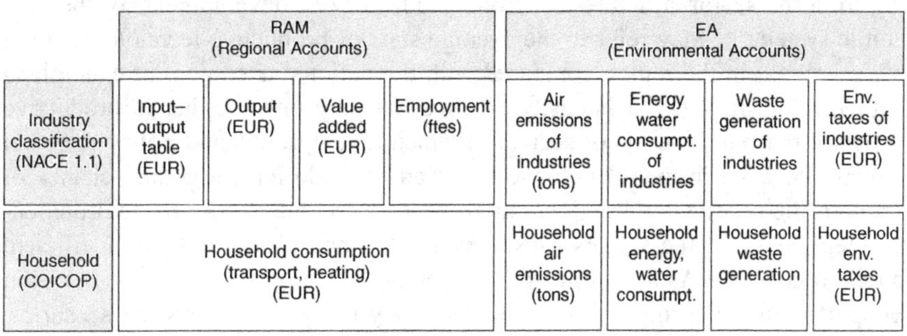

Figure 4.1 RAMEA framework (Arpa Emilia-Romagna).

for effects and factors that explain the relative eco-efficiency of Emilia-Romagna, compared with Italy, which could be shown in a more exhaustive way than a descriptive statistic analysis (the differences between the regional and national indicators of intensity of emission). The deviation matrix between the regional and national average, generated by a descriptive statistic analysis, can be investigated by application of shift-share analysis to carry out detailed considerations on these differentials (Table 4.1). In this study, the total differentials of efficiency for GHG do not remain in favour of Emilia-Romagna for every sector since otherwise there would be an advantage for the whole regional economic system when compared with the national one. It is explained by both an industry mix effect and a differential one.

The regional average intensity of emission (X_e) for GHG is the summation of sector intensity of emission (X^s_e), weighted for the sector ratios of the total value added (P_e). The national average intensity (X) is defined in the same way. The region can show a total higher or lower intensity of emission compared with the national average caused by the combination of the three shift-share effects: Industry mix (m_e), Differential (p_e), Allocative (a_e). The total difference between regional and national average intensity of emission equals the summation of the three effects ($X_e - X = m_e + p_e + a_e$). The Industry mix estimates the part of higher/lower intensity of emissions that is due to the sector structure of the economic system. The difference between regional and national average intensity of emission could depend on differences in the specific intensity of emissions of some or all considered fields marking out the Differential effect. Finally, the Allocative component adds further analytic information: the covariance between sector structure (assuming parity of efficiency) and difference between sector intensity of emission (assuming parity of sector structure) indicates how much and if the system has a productive specialization in the fields where it carries out a comparative advantage of efficiency. This is reflected in the interpretation of differential between Emilia-Romagna and Italy; therefore, if $X_e - X > 0$, Emilia-Romagna is relatively less efficient (i.e. produces more emissions for unit of value added than the national average). The same is true for the signs of the three effects: when they are algebraically negative, they mark an advantage of efficiency for the Emilia-Romagna region. These effects show influences deriving from the sector structure and from 'the history of development' of the economic system, or may refer to the average state of productive technologies (and of emissions) in the region compared with the national average. For example, a higher value of regional intensity of emission may only be due to productive structure reasons in terms of sectors in which an environmental policy does not directly have a substantial influence; instead, it could have a greater chance of action if the relative total regional inefficiency were due to specific environmental inefficiency of the sectors caused by the technologies used or by inefficient public regulation. As a result of this reasoning and processing, a pilot Decision Support Matrix is proposed as an aid to policy-makers: it shows the scenarios, depending on the possible combination of shift-share effects and identifies strategies for sector policy.

Table 4.1 Shift-share analysis applied in Emilia Romagna: simplified matrix 2000. (Mg CO_2eq/Meuro); $X_e - X = m_e + p_e + a_e$

GHG	$\sum X_e$	$\sum X$	$\sum (X_e - X)$	$\sum (m_e + p_e + a_e)$	$\sum m_e$	$\sum p_e$	$\sum a_e$
Total economic activities	0.3405	0.4129	−0.0724	−0.0724	−0.0248	−0.0753	0.0276
Sectors	X_e^s	X^s	$(X_e^s * P_e^s) - (X^s * P^s)$	$m^s + p^s + a^s$	m^s	p^s	a^s
A+B	1.6036	1.6926	0.0091	0.0091	0.0122	−0.0025	−0.0006
C	0.2843	0.1347	−0.0002	−0.0002	−0.0004	0.0007	−0.0005
D	0.5483	0.4914	0.0440	0.0440	0.0288	0.0118	0.0033
E	5.0696	9.0571	−0.1125	−0.1125	−0.0603	−0.0787	0.0265
F	0.0149	0.0492	−0.0017	−0.0017	−0.0000	−0.0017	0.0000
G+H	0.0531	0.1169	−0.0104	−0.0104	0.0006	−0.0106	−0.0003
I	0.4763	0.5165	−0.0054	−0.0054	−0.0026	−0.0029	0.0002
J–Q	0.0731	0.0538	0.0047	0.0047	−0.0029	0.0087	−0.0010

Source: own calculations.

Extension to new environmental issues

The idea for an extension is based on the ESEA (Eurostat, 2008) and the European Commission proposal on environmental economic accounts (2010) which suggest developing data collection in the areas of energy, waste accounts and environmental taxes. The release of 2005 regional NAMEA air emission accounts for Italian regions by ISTAT (Italian National Statistics) is thus taken as a robust base for studying the opportunities of extending the framework to new environmental issues: (a) environmental taxation scheme, by downscaling national statistics data; this research also investigates the use of eco taxes (which have long been an instrument used to boost the behaviour change of citizens by giving monetary values to negative externalities on the environment) coordinated with RAMEA; (b) energy consumptions of industries and households by processing regional data provided by ENEA (Italian National Agency for Energy) and TERNA SpA (Italian company responsible for electricity transmission); (c) waste production of industries, using regional datasets by Arpa Emilia-Romagna.

The extension methodology follows two main steps, applied to the link between the RAMEA (which uses NACE and COICOP codes) classification system and the other available datasets: (a) the analysis of the qualitative link; (b) the quantitative allocation of data.

While with the first step we can make a distinction between *one-to-one* and *one-to-several* correspondence (between environmental data and RAMEA categories), in the second one, we have to quantitatively assign data from environmental issues to RAMEA categories. In the *one-to-one* correspondence, the allocation is easy (100 per cent of the environmental data goes to the corresponding RAMEA category) whereas in the *one to several* link, the use of proxy variables (e.g. value added, CO_2 emissions) is fundamental for splitting the value of environmental data between the several RAMEA categories.

Environmental taxes

Following EU directions, environmental taxes[5] have long been an instrument for boosting the behaviour change of citizens by giving monetary values to negative externalities on the environment, such as polluting, and also by increasing the costs of certain products which have a negative impact on the environment and an instrument for adjusting revenues in national budget spending or reducing other taxes. European efforts, such as the Lisbon Strategy, emphasize that environmental taxes are an important tool, not only for the protection of the environment but also for competitiveness and growing economies. The green tax reform[6] should lead to decreasing labour taxes and more weight being placed on environmental taxes. Environmentally related taxes can often be usefully implemented in the context of instrument mixes in combination with other policy tools such as command and control regulations, tradable permits, voluntary approaches and environmental accounting tools. Of the environmental policy tools, environmental taxes are considered to be environmentally effective and economically efficient. The OECD has supported the use of these tools and has carried out an analysis of their implementation (2001). The Sixth Community Action Programme on the Environment, approved in 2002, and the European Commission's Green Paper (2007) recommend the use of economic instruments (energy taxes, taxes on resources) to mitigate climate change and promote sustainable use of resources.

Regional data for environmental taxes are not available in Italy yet, but Eurostat and ISTAT provide environmental taxes split up into economic activities and household consumption (following NACE classification) at national scale.[7] In Italy, three kinds of environmental taxes are now available: energy taxes, pollution taxes and transport taxes. In particular, the CO_2 taxes are included under energy taxes rather than under pollution taxes, and the second one includes taxes on measured or estimated emission to air and water and management of solid waste.[8]

In order to build a RAMEA matrix integrated with eco-taxes, regional eco-taxes had to be estimated by downscaling the national data: value added and household consumptions were identified as good proxies to perform the analysis.

A very good statistical correlation (0.94) was obtained between total regional and national values added (historical series 2000–2006) and an excellent correlation (0.96) between regional and national household final consumption (Bonazzi *et al.*, 2009).

Using the above findings, the three environmental taxes available at national level (energy, pollution, transport) were downscaled at regional level and split up in economic activities and household using the following formulas:

$$ET_{ER,i} = \frac{VA_{ER,i}}{VA_{IT,i}} \cdot ET_{IT,i} \tag{1}$$

for economic activities, where $ET_{ER,I}$ is the regional environmental tax for the *i-th* sector, $VA_{ER,i}$ is the regional value added for the *i-th* sector, $VA_{IT,i}$ is the national value added for the *i-th* sector and $ET_{IT,i}$ is the national environmental tax for the *i-th* sector and

$$ET_{ER,H} = \frac{H_{ER}}{H_{IT}} \cdot ET_{IT,H} \tag{2}$$

for household, where $ET_{ER,H}$ is the regional environmental tax for household, H_{ER} is the regional household consumption, H_{IT} is the national household consumption and $ET_{IT,H}$ is the national environmental tax for households.

As explained above, thanks to the high statistical correlation verified among regional and national value added and household consumptions, we downscaled eco-taxes at regional level referring to these proxies. So it is important to remark that the outcomes are estimations. Therefore, attention when evaluating the quality of numbers obtained is strongly recommended; in this case the attempt is essentially to show the structure of an environmental economic tool, on a regional scale, useful for monitoring, controlling and addressing the effects of environmental fiscal policies. It is important to take care of the structure proposed and the relevance in addressing statistical offices to provide data at local scale in order to support sustainable local policies.

Industrial waste production

Some interesting experiences already exist in Italy concerning the application of the NAMEA approach to industrial waste production (Dalmazzone and La Notte, 2009). The integration of RAMEA with the production of industrial waste is effected using data provided by the regional thematic centre for integrated management of waste (Arpa Emilia-Romagna). Data on industrial waste production is collected annually from waste producers using specific surveys (MUD – declaration of industrial waste production) and collected in a regional database (waste cadastre). Even if the data collected by MUD has some limitations (the use of MUD could lead to an underestimation of the actual amount of total waste generated, taking into account that not all manufacturers are obliged to submit the MUD declaration, not all types of waste have to be declared and a number of individuals do not fulfil the obligation to submit the declaration), it should be noted that they are to be considered official data. Thanks to the availability of production of industrial waste already divided into NACE codes, the processing of data for RAMEA purposes is easier than for other environmental issues and all correspondences fall into the *one-to-one* category.

Energy consumption

Integration with the energy consumption issue is performed using data from regional energy balances (provided by ENEA for overall energy consumptions in toe – tons of oil equivalent – and by TERNA SpA for electricity consumptions in GWh). TERNA is the electricity transmission system operator and annually analyses electricity consumption according to different types of users at national, regional and provincial level. The TERNA classification is a modified version of NACE classification and therefore *one-to-one* correspondences to RAMEA sectors (and electricity consumptions in GWh) can be easily found. ENEA is the Italian National Agency for New Technologies, Energy and Sustainable Economic Development and, among its other duties, is responsible for drawing up national and regional energy balances. The energy balance contains information on supply, transformation and use of different energy sources (solid, liquid and gaseous fuels, renewables and electricity in toe) in four macro-sectors (Agriculture and fishing, Industry, Residential and Transportation). Each macro-sector is then divided into sub-sectors. The ENEA classification slightly differs from NACE in that most of the correspondences are *one-to-one*, but some *one-to-several* can be found.

First, the ENEA Textile sector energy consumption is split into DB and DC NACE codes using the following formulas:

$$EN_{DB} = \frac{VA_{DB}}{(VA_{DB} + VA_{DC})} \cdot EN_{textile} \tag{3}$$

$$EN_{DC} = \frac{VA_{DC}}{(VA_{DB} + VA_{DC})} \cdot EN_{textile} \tag{4}$$

where EN_{DB} and EN_{DC} are the energy consumption in toe for DB and DC regional sectors, VA_{DB} and VA_{DC} are the regional value added for DB and DC sectors and $EN_{textile}$ is the energy consumption in toe of the Textile sector as reported in the ENEA regional energy balance.

Second, energy consumptions from ENEA sector Transportation has to be distributed to all NACE and COICOP activities (and summed to the energy consumption already distributed). In a first estimation, RAMEA data on CO_2 emitted by each NACE and COICOP codes are selected as a proxy variable to distribute energy consumption using the following formula:

$$EN_{transportation,i} = \frac{CO_{2i}}{CO_2} \cdot EN_{transportation} \tag{5}$$

where $EN_{transportation,i}$ is the quantity of energy consumption in toe from ENEA Transportation sector to be distributed to the RAMEA *i-th* sector (when *i* covers all NACE and COICOP codes used), CO_{2i} is the tons of CO_2 emitted by the *i-th* sector of RAMEA, CO_2 is the total emission from all RAMEA sectors and $EN_{transportation}$ is the energy consumption in toe of the Transportation sector as reported in the ENEA regional energy balance.

Using the above assumptions, a first draft of RAMEA air emissions extended to environmental taxes, industrial waste production and energy (electricity plus total energy) consumptions can be drawn up (Table 4.2). The structure of RAMEA can be used to identify the different contributions of economic sectors and households to the economy and the environment. If values are calculated as a percentage of the total, it becomes immediately obvious how much each sector contributes to the economy and to environmental pressures respectively (Figure 4.2).

Conclusion

In the long term, the overall aim of this study is to outline RAMEA as a policy tool to support sustainable regional policies and possibly environmental assessments of regional plans and programmes (the most important being the Regional Territorial Plan). In this context, the opportunity of building a scenario analysis based on the extended RAMEA should be studied and regional sustainable development steered by means of a more complete environmental accounting system.

The pilot extended RAMEA needs to be updated, using the most appropriate economic and environmental data sources: the air emission accounts issue, in particular, could be built using data coming from the regional air emissions

Table 4.2 RAMEA air emissions extended to eco-taxes, ind. waste production and energy consumptions (2005)

RAMEA 2005 Emilia-Romagna (NACE/COICOP code)	*Regional accounts*			*Environmental accounts*									
	VA	*FIN CONS*	*FTE*	*GHG*	*ACID*	*NO_x*	*PM_{10}*	*ETAX – EN*	*ETAX – POLL*	*ETAX – TR*	*IND WASTE*	*ELECTRICITY*	*ENERGY*
	(MEur 2000)	*(MEur 2000)*	*(thousands of units in average)*	*(Mtons of CO_2 eq)*	*(tons of acidification potential)*	*(tons)*	*(tons)*	*(MEUR)*	*(MEUR)*	*(MEUR)*	*(tons)*	*(GWh)*	*(toe)*
COICOP 07		**8,806.94**		4,248.67	324.77	11,726.00	1,079.25					–	988,727.64
COICOP 04		**11,822.00**		7,706.50	188.26	6,810.70	877.87					5,038.70	3,175,285.32
COICOP total		**42,967.43**		45.74	–	–	–					–	–
Household – total		**63,598.80**		**12,000.91**	**513.03**	**18,536.70**	**1,957.12**	**1,149.90**	**5.47**	**323.62**		**5,038.70**	**4,164,012.96**
A	2,954.17		109.40	5,259.60	3,204.00	9,032.90	4,873.46	57.73	0.33	8.56	118,936.96	919.10	535,649.34
B	51.92		3.90	51.14	15.01	688.37	72.61	3.17	0.01	0.18	29.39	–	33,975.40
C	145.58		1.60	335.62	16.22	367.60	66.49	2.92	0.04	0.15	115,414.77	70.60	50,094.35
DA	3,555.18		71.90	2,972.24	131.83	3,296.20	378.54	64.34	1.76	4.26	1,209,144.69	2,441.10	1,413,769.71
DB	1,616.14		47.00	486.74	14.67	617.29	29.88	27.74	0.79	2.67	10,313.39	237.40	112,867.98
DC	323.65		9.90	97.19	4.19	119.18	15.33	2.64	0.18	0.63	18,235.83	43.60	21,679.75
DD-DH-DN	2,074.82		51.10	533.20	22.49	819.54	74.78	17.99	0.91	3.03	481,335.76	1,387.30	252,022.44
DE	1,024.81		22.10	433.38	8.57	382.11	20.28	17.70	0.52	1.02	199,710.06	662.00	269,135.60
DF-DG	1,415.20		15.90	5,465.36	248.18	4,587.40	405.63	52.42	1.54	0.85	171,176.47	1,590.90	869,265.64
DI	2,679.19		47.50	6,077.85	678.01	10,829.00	2,242.22	159.04	1.20	3.47	1,221,369.16	3,188.80	2,729,938.23
DJ	3,998.02		91.90	258.35	18.53	769.63	562.83	87.96	2.07	4.73	516,658.32	206.30	72,689.75
DK-DL-DM	8,558.30		174.60	1,327.89	51.92	2,137.40	138.49	68.31	2.17	7.87	283,373.00	3,157.10	783,361.36

E	1,799.45		9.90	7,039.40	91.80	3,418.30	96.14	194.14	3.09	0.81	227,167.43	571.20	5,004,576.40
F	5,434.36		147.10	330.47	32.32	1,378.10	366.66	62.72	0.97	13.67	98,726.21	223.20	63,562.59
G	11,226.76		291.70	1,637.25	158.45	7,187.40	678.03	185.16	4.87	25.87	590,704.81	1,932.00	826,386.70
H	3,424.79		131.50	280.94	23.71	1,070.20	77.69	32.58	1.40	2.30	6,918.01	1,079.50	385,695.98
I	6,561.64		128.00	2,452.74	369.85	13,605.00	1,244.93	240.01	2.34	7.95	174,578.89	1,163.10	294,419.84
J	4,755.94		51.00	100.38	8.81	389.24	34.31	12.02	1.50	1.64	573.03	244.40	91,794.90
K	20,177.02		231.60	626.94	62.97	2,801.30	258.79	76.77	1.86	9.11	58,864.95	97.03	106,789.03
L	3,676.74		77.50	248.15	32.87	1,331.10	131.71	19.98	1.49	1.42	27,940.64	684.40	156,435.76
M	3,185.17		93.50	75.09	4.25	158.78	8.61	7.95	0.01	0.58	112.44	185.81	69,723.38
N	5,221.89		128.10	333.05	14.31	586.09	30.90	23.64	0.85	2.62	21,679.87	155.97	85,392.07
O	2,420.54		90.20	1,935.19	96.39	1,893.40	135.57	23.39	3.95	2.11	3,118,512.01	1,341.30	518,631.09
P-Q	844.16		57.70	–	–	–	–	–	–	–	129.20		
Economic activities – total	**97,174.29**		**2,084.60**	**38,358.15**	**5,309.36**	**67,465.53**	**11,943.87**	**1,440.32**	**33.86**	**105.51**	**8,671,605.29**	**21,582.11**	**14,747,857.29**
TOTAL	**97,174.29**	**63,598.80**	**2,084.60**	**50,359.06**	**5,822.39**	**86,002.23**	**13,900.99**	**2,590.22**	**39.33**	**429.13**	**8,671,605.29**	**26,620.81**	**18,911,870.25**

Legend

VA: value added basic prices (source: ISTAT); FIN CONS: final consumption (source: ISTAT); FTE: full-time equivalents (source: ISTAT); GHG: greenhouse gases (source: ISTAT); ACID: acidification (source: ISTAT); NO_x: nitrous oxides (source: ISTAT); PM_{10}: particulate matter (source: ISTAT); ETAX – EN: environmental taxes – energy (source: own calculations on Eurostat data); ETAX – POLL: environmental taxes – pollution (source: own calculations on Eurostat data); ETAX – TR: environmental taxes – transport (source: own calculations on Eurostat data); IND WASTE: industrial waste (source: Arpa Emilia-Romagna); ELECTRICITY: electricity consumptions (source: own calculations on TERNA SpA data); ENERGY: total energy consumptions (source: own calculations on ENEA data).

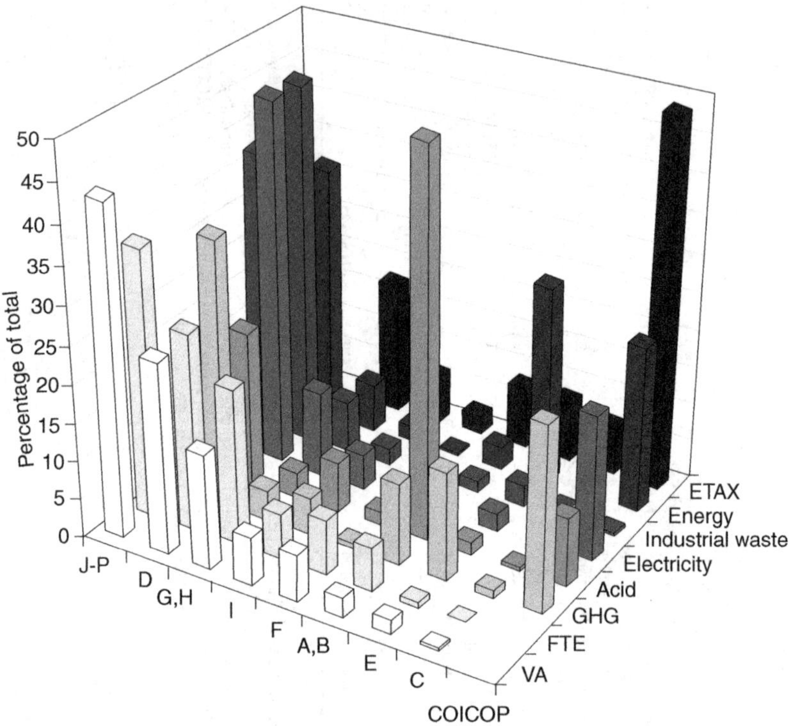

Figure 4.2 Contribution of different sectors to the economy and the environment, (2005, %).

inventory by Arpa Emilia-Romagna and following Eurostat guidelines. The availability of up-to-date RAMEA matrices will enhance the opportunity to study the integrated economic-environmental performances of the region and possibly answer important policy questions, as highlighted by Eurostat (2009).

Summing up, RAMEA framework, in its most recent comprehensive version and in its future developments, could be regarded in the Emilia-Romagna region as: (a) a *monitoring* system which analyses the pressure placed on the environment by the economic sectors and households, helps to identify the 'hot spots' in terms of environmental pressures and potential decoupling patterns, allows processing of eco-efficiency indexes, uses the knowledge base on the economic and environmental performances of regional sectors and enforces the role of policy tools in promoting sustainable behaviour (e.g. regarding eco-taxes 'to make the polluter pay'); (b) a tool that allows *scenario* analysis to evaluate the economic-environmental effects of the policies; (c) a *benchmarking* tool that can be used to compare European regions and countries; (d) an *evaluation* tool that helps to assess policy effects on the economic system, identify which are the most efficient (eco-efficient) sectors in the region and, together with an input–output matrix, could be helpful in verifying environmental-economic inter-relations between the sectors; (e) a *motivating* tool that should strengthen the final goal of environmental taxes by creating incentives for producers and consumers to move away from environmentally damaging behaviour; thanks to RAMEA, environmental taxes could also be applied more efficiently in the long term, by acting in proper economic sectors.

Acknowledgements

The authors are particularly grateful for their suggestions and data provided to: Paolo Acciari (Ministero dell' Economia e delle Finanze), Cecilia Cavazzuti, Barbara Villani, Giacomo Zaccanti (Arpa Emilia-Romagna), Massimiliano Mazzanti (University of Ferrara), Anna Montini (University of Bologna), Martina Ruffilli (University of Bologna student), Angelica Tudini (Istat). All errors are our own.

Notes

1 See Kennedy (1968), Hall (2005), Matthews (2006), Almunia (2007), Commission of the European Communities (2009), Stiglitz *et al.* (2009), the Beyond GDP International Conference (2007; www.beyond-gdp.eu) and the Global Project on Measuring the Progress of Societies – OECD (2008; www.wikiprogress.org/index.php/Global_Project).

2 Communication COM (94) 670 (Commission of the European Communities, 1994).

3 The application in Emilia-Romagna benefited from previous pilots of regional NAMEA for two Italian regions, Toscana (by IRPET – Tuscan Regional Institute for Economic Planning) and Lazio (by ISTAT – Italian National Statistics), together with the compilation of national and regional NAMEA for Italy by ISTAT.

4 The term *eco-efficiency* was coined by the World Business Council for Sustainable Development (WBCSD) in its 1992 publication 'Changing Course'. It is based on the concept of creating more goods and services while using fewer resources and creating less waste and pollution. Following this indicator, the quantity of emissions is related to the gross value added of an economic sector. The lower the indicator, the more eco-efficient the sector; additionally, regional eco-efficiency can be compared with the overall eco-efficiency of the country or with other national or foreign regions. Policy-makers are able to use these indicators to find which sectors of the economy should increase their eco-efficiency.

5 Eco-taxes are those whose tax base has a proved harmful effect on the environment, e.g. a process or product which pollutes the environment. The aim of environmental taxes is to internalize external environmental costs by focusing on limitation of environmental burden and responsible use of natural resources by producers as well as consumers. They are divided into the following categories: energy taxes, transport taxes, pollution taxes, resource taxes.

6 www.eea.europa.eu/highlights/green-tax-reform-can-boost-eco-innovation-and employm ent.

7 http://epp.eurostat.ec.europa.eu/portal/page/portal/environmental_accounts/data/database.

8 See Eurostat (2001).

References

Almunia, J. (2007) 'Measuring progress, true wealth and well being', speech delivered at Beyond GDP International Conference, Brussels: European Commission.

Bonazzi, E. and Sansoni, M. (2008) 'Evaluation of the level of green house gas emissions in Emila-Romagna region: a statistical shift share analysis to develop the decision support systems', *Valutazione Ambientale*, 13: 18–25.

Bonazzi, E., Goralczyk, M., Sansoni, M. and Stauvermann, P.J. (2008) 'RAMEA: A decision support system for regional sustainable development', 14th Annual International Sustainable Development Research Conference, Conference Proceedings, New Delhi, 21–23 September.

Bonazzi, E., Sansoni, M., Setti, M., Cagnoli, P. and Bontempi, S. (2009) 'RAMEA, a shared environmental accounting tool to control and monitor regional environmental taxes', 10th Global Conference on Environmental Taxation, Conference Proceedings, Lisbon, 23–25 September, p. 51.

Commission of the European Communities (1994) 'Directions for the EU on environmental indicators and green national accounting: the integration of environmental and economic', *Information Systems*, COM (94) 670, Brussels: European Commission.

Commission of the European Communities (2007) 'Green Paper on market-based instruments for environment and related policy purposes', COM (2007) 140 final, Brussels: European Commission.

Commission of the European Communities (2009) 'Beyond GDP: measuring progress in a changing world', COM (2009) 433 final, Brussels: European Commission.

Dalmazzone, S. and La Notte, A. (2009) 'The NAMEA approach for air emissions and wastes applied at regional, provincial and municipal level', *Economics and Policy of Energy and the Environment*, 3: 61–86.

Dosi, M.P., Bonazzi, E. and Sansoni, M. (2008) 'Progettare la sostenibilità nello sviluppo di un territorio: l'analisi shift share su aggregati economico-ambientali', XXIX Conferenza Italiana di Scienze Regionali AISRE, Conference Proceedings, Bari.

European Commission (2008) 'Regions and cities in a challenging world', Open Days 2008 Proceedings, Brussels: European Commission.

European Commission (2010) 'Proposal for a regulation of the European Parliament and of the Council on European Environmental Economic Accounts', COM (2010) 132 final, Brussels: European Commission.

Eurostat (2001) *Environmental Taxes: A Statistical Guide*, Brussels: European Commission.

Eurostat (2008) 'Revised european strategy for environmental accounting', 68th meeting of the Statistical Programme Committee, E-3 CPS 2008/68/7/EN.

Eurostat (2009) *Manual for Air Emissions Accounts*, Methodologies and Working Papers: Environment and Energy, Brussels: European Commission.

Goralczyk, M. and Stauvermann, P.J. (2007) 'The usefulness of hybrid accounting systems for environmental policy', Advice Regarding Sustainability, 16th International Input–Output Conference, Istanbul.

Hall, J. (2005) 'Measuring progress: an Australian travelogue', *Journal of Official Statistics*, 21.

Kennedy, R.F. (1968) 'Speech at University of Kansas', mimeo, 18 March.

Matthews, E. (2006) 'Measuring well-being and societal progress: a brief history and the latest news', OECD-JRC Workshop, Milan.

Mazzanti, M. and Montini, A. (2009) 'Regional and sector environmental efficiency:

empirical evidence from structural shift-share analysis of NAMEA data', FEEM Working Paper No. 11, Milan: FEEM.

Mazzanti, M., Montini, A. and Zoboli, R. (2007) 'Struttura produttiva territoriale ed indicatori di efficienza ambientale attraverso la NAMEA regionale: Il caso del Lazio', *Economia delle Fonti di Energia e dell'Ambiente*, 40.

OECD (2001) 'OECD environmental strategy for the first decade of the 21st century', Meeting of the Environment Policy Committee at Ministerial Level: ENV/EPOC (2000) 13/REV4, Paris: OECD.

RAMEA (2007) 'Construction manual, user manual and case studies', INTERREG IIIC GROW RAMEA project report. Online: www.ramea.eu.

Sansoni, M., Bonazzi, E., Goralczyk, M. and Stauvermann, P.J. (2010) 'RAMEA: how to support regional policies towards sustainable development', *Sustainable Development*, 18: 201–210.

Stiglitz, J.E., Sen, A. and Fitoussi, J. (2009) 'Report by the Commission on the Measurement of Economic Performance and Social Progress'. Online: www.stiglitz-sen-fitoussi.fr.

5 Feasibility and uses of the NAMEA-type framework applied at local level

Case studies in North-Western Italy

Alessandra La Notte and Silvana Dalmazzone

Introduction

The National Accounting Matrix including Environmental Accounts (NAMEA) is currently applied at national level in most industrialized countries in compliance with the SEEA 2003 (*Handbook of National Accounting: Integrated Environmental and Economic Accounting* – UN, 2003) international standard. Environmental accounting at local government level cannot yet rely on a comparably established methodology. Its development is to be tracked to the Local Agenda 21 process and its aim tends to point to the practical understanding of environmental information for communicating the objectives and the results of local policies rather than to the suitability of the statistical standard for strategic planning and policy-making. This bottom-up approach has generated a number of different green budgeting schemes in different countries, the most important of which is probably the Eco-Budget method, developed by the International Council for Local Environmental Initiatives (ICLEI), a worldwide network of local government units that supports sustainable development initiatives. It consists of a budgeting system for natural resources that conforms to the existing financial budgeting procedures of local government and is based on environmental indicators that offer aggregate information already processed with the purpose of providing insights into the trend of given environmental issues or the effectiveness of specific policies (e.g. ambient concentration of a given pollutant, the quantity of municipal waste per inhabitant, the average age of circulating motor vehicles, and so on). The focus is typically on the municipality or local jurisdiction, the indicators to be included are subjectively defined according to the interests of local authorities and the temporal span generally includes a few sample years so that identification of the direction of the trend is possible.

At local level, the implementation of integrated environmental and economic accounts – that is, systems based on proper accounting schemes of the kind standardized in the SEEA approach, with a regular, long-term structure that integrates the pre-existing standard economic accounting – is still experimental and until now has only been experienced occasionally, in isolated contexts. It appears, however, to be the natural direction for the development of organic environmental accounts harmonized between national and sub-national levels. Policies for many environmental and natural resources are, in most countries, designed and implemented at sub-central levels of government. Municipalities, for instance, are often in charge of urban pollution-control policies and the control of land use and protected areas is often assigned at intermediate levels (regional/provincial). Making environmental accounting an operational tool not only for reporting results, but also for setting objectives and designing policies, requires detailed accounts of a kind that allows analysts to trace the origin of the emission or resource consumption by sector and sub-sector of economic activity. This ought to be compiled not only on a national but also on a local scale, and thus made available to local planners and decision-makers who have the responsibility of administering and regulating natural resources, local development actions and conservation policies (Dalmazzone and La Notte, 2009).

There is virtually still no experience of a comprehensive system of environmental accounting systematically extended to all levels of government. In several countries, however, environmental accounting modules based on some kind of rigorous accounting framework have been tested at local level. In Italy, the national statistical office (ISTAT) has tested the compilation of NAMEA for air emissions in all Italian regions.[1] Within the European project GROW, five European regions have started to compile NAMEAs for air emissions. In the Netherlands, some applications related to water accounts have been implemented with respect to river basins. In Sweden, there has been work on regional accounts for the Stockholm area and several studies have been made at district level for water accounts. In Germany, the statistical offices of the Länders compile material flow accounts at regional level, and generally the results for the 16 Länders sum up to the results for Germany as a whole. In Canada, there are two NGOs producing environmental accounts at provincial level, but there is no coordination activity between the national statistical office and these organizations. The same occurs in the United Kingdom where a local NGO produces environmental accounts for Wales. In New Zealand, physical stock accounts for water are compiled by the central statistical office on both a national and a regional scale. In the Philippines, in the Cordillera Administrative Region and in the province of Palawan, all asset accounts compiled at national level are also compiled at regional and provincial levels.

There is no comprehensive research to date that compares the experiences of these different experimental applications and draws lessons and methodological implications. In this chapter, we present indications stemming from pilot applications of NAMEA accounts in the Piedmont Region (Italy) at regional, provincial and municipal level conceived with the specific purpose of testing the feasibility and reliability of integrated environmental and economic accounts at all sub-national levels of government. The aim is to show how to apply NAMEA at different scales, identifying criticalities and possible solutions, and checking to what extent the obtained results prove to be useful.[2]

General features of a local NAMEA-type account

In local environmental accounting, the data for filling both the economic and environmental components of NAMEA should be gathered locally: using proxies calculated at national level fails to serve the very purposes of compiling local environmental accounts. Data should in fact reflect the peculiarity of local contexts, and this does not happen when specific local values are averaged on a larger scale. The first step is therefore to investigate whether there are datasets at local level and whether they are suitable to be used for this purpose.

Local governments and agencies usually maintain accurate and in-depth environmental databases with locally gathered, bottom-up information. As far as economic data are concerned, national statistical offices generally produce statistics at national and sometimes regional levels. The only statistical data that systematically reach the level of municipalities are census surveys which usually take place only every ten years. However, data at sub-regional levels can be collected by integrating statistical databases with local administrative archives and other minor sectoral and business registers.

Even when available, local data must be precise, accurate and homogeneous to be useful. What should be known and checked is their sources,[3] processing procedure,[4] scale[5] and timing.[6] The methodological issues that arise when compiling the NAMEA-type framework at local level concern the lack of data at detailed territorial level and the fact that data may not be classified in an appropriate way to fill the accounting module.

In the case studies presented in the following sections, NAMEA for air emissions is compiled at regional and sub-regional levels and refers to the year 2005.

In Italy, economic data can be obtained from the register of active enterprises (named ASIA). The sources of ASIA are both statistical (derived from ISTAT) and administrative data (derived by chambers of commerce, the institute for social security, the revenue office, telephone subscriptions, banking and insurance institutes, and so on). The integration and harmonization of statistical and administrative data is undertaken by ISTAT in order to accomplish what is required by European legislation (Council Regulation (EEC) No. 2186/93). Thanks to ASIA, we can retrieve data on the number of employees and local units at regional, provincial and municipal level according to the NACE (Nomenclature générale des Activités économiques dans les Communautés Européennes, Rev. 1.1) classification from 2004 to 2007. However, data on value added and production are not available at present from ASIA although ways of calculating them using the ASIA input data are being tested (Faramondi, 2008). Primary and public sectors are also not currently included, but their inclusion is planned for the near future.

Environmental data can be acquired from the Regional Inventory of Air Emissions (IREA) maintained by the regional Environmental Protection Agency (EPA). Data are estimated according to the CORe INventory AIR emissions (CORINAIR) method, the framework supported by the European Environment Agency and adopted by the national environmental protection agency to compile the national inventory. IREA, however, retains a bottom-up focus: all the information on economic activities refers to a municipality or a geographically identified location. IREA records data according to the SNAP (Selected Nomenclature for Air Pollution) classification.[7] This poses a reclassification problem: the SNAP process classification must be turned into NACE sector classification and emissions generated by natural processes have to be excluded. Reclassification implies a qualitative assignment when a correspondent SNAP production process is assigned to each NACE economic activity, and a quantitative assignment whenever the link to the process involves multiple economic activities and emission allocations must be estimated.[8] In both qualitative and quantitative assignments, it is important to know how emissions were estimated for each SNAP process. To this end, cooperation must be established between the environmental accounting analyst and the statisticians or technical staff that built the inventory and have first-hand knowledge of territorial characteristics such as land use and the distribution of productive settlements.

Once the emissions have been attributed to each NACE activity, three indicators are calculated, as suggested in the NAMEA handbook (European Commission, 2009): greenhouse gases (based on 'global warming potential'),[9] acidification[10] and tropospheric ozone (based on tropospericozone formation potential).[11]

In the next section, we are going to present an application of NAMEA at regional and sub-regional levels and discuss the information content that this accounting module can provide to local policy-makers.

Case studies: descriptive analysis

A comparison between two main northern regions: Lombardia and Piemonte

The first application of NAMEA is tested at regional and provincial levels in two bordering regions in North-Western Italy, Piemonte and Lombardia. The application is then used to undertake comparisons at both horizontal level (i.e. between the two regions) and vertical level (i.e. between the region and one of its provinces). The two regions are characterized by well developed industrial sectors: in Piemonte, the automotive industry (the Fiat group and its induced activities) is the dominating compartment followed by chemical, food, textile, clothing, electronics and editorial compartments. The tertiary sector is also well developed with banking, insurance, trade and tourism. In Lombardia, the secondary sector is important in the mechanics, electronics, metallurgic, textile, chemical and petrochemical, pharmaceutical, food, shoes and furniture compartments whereas trade and finance dominate the tertiary sector.

The two regions show very similar characteristics and both host important capitals (Turin and Milan) where most of the population is concentrated. The data sources utilized for air emissions are IREA for the Piemonte region and INEMAR for the Lombardia region. Once data on pollutant emissions have been properly reclassified, the NAMEA-type framework is compiled and the greenhouse gases, acidification and tropospheric ozone indicators calculated. Table 5.1 shows the impact of each economic macro-sector in terms of each pollutant and the three aggregate environmental indicators.

Table 5.1 Impact of air emissions by macro-sectors in Piemonte and Lombardia, 2005 (%)

	CH_4 (t)	*CO (t)*	*CO_2 (kt)*	*N_2O (t)*	*NH_3 (t)*	*NMVOC (t)*	*NO_x (t)*	*PM_{10} (t)*	*SO_2 (t)*	*Greenhouse gases*	*Acidification*	*Tropospheric ozone*
PIEMONTE REGION												
Agriculture	63.29	3.79	2.73	44.36	95.04	2.81	13.31	22.32	1.19	11.54	51.71	9.36
Industry	10.63	16.97	47.86	45.79	0.61	43.38	30.63	27.11	78.67	43.46	21.02	33.55
Services	24.83	10.15	7.29	3.41	1.25	15.52	20.22	11.36	3.05	9.15	9.25	17.58
Households	1.25	69.09	42.12	6.45	3.11	38.29	35.83	39.20	17.09	35.84	18.03	39.51
LOMBARDIA REGION												
Agriculture	*47.75*	3.22	1.77	71.62	96.77	1.37	8.40	16.79	1.34	9.89	55.35	5.13
Industry	24.58	12.81	48.23	8.71	0.38	41.54	33.34	21.39	81.53	43.86	19.79	35.25
Services	25.18	10.72	11.59	4.89	0.35	7.54	31.93	14.52	2.89	12.58	12.41	18.63
Households	2.49	73.25	38.42	14.77	2.50	49.55	26.34	47.31	14.24	33.67	12.46	40.98

Some sectors have an almost identical impact in the two regions, but for some forms of pollution – consider, for instance, nitrous oxide (N_2O), non-methane volatile organic compounds (NMVOC) and nitrogen oxides (NO_x) – there also appear to be important differences. In order to understand why, the macro-sector (industry) must be broken down into more detailed economic activities and the origin of each pollutant traced.

N_2O and NMVOC emissions in Piemonte are generated by the activities 'Manufacture of other non-metallic mineral processing', 'Manufacture of pulp, paper and paper products, printing and publishing' and 'Manufacture of wood, rubber, plastics and other manufacturing', respectively. NO_x emissions in Lombardia are generated by the activities 'Manufacture of basic metals and fabricated metal products' and 'Production and distribution of electricity, gas, steam and water' (Table 5.2).

NO_x emissions in Lombardia are generated by the 'Manufacture of basic metals and fabricated metal products' and 'Production and distribution of electricity, gas, steam and water' activities (Table 5.3).

However, if we consider data in absolute terms and not in terms of their percentage impact, Lombardia records emissions which are almost double those in Piemonte (Figure 5.1). The number of jobs in the secondary and tertiary sectors (thus excluding the public sector) in Lombardia is 4,013,655 units, whereas it is 1,434,764 units in Piemonte.

A comparison between the Turin and Milan provinces and their respective region reveals important differences (vertical comparison in Table 5.1 and Table 5.3), and so does a comparison between the two provinces (horizontal comparison in Table 5.3).

Contrary to what occurs at regional Piemontese level, in the province of Turin, the primary sector is not the biggest methane polluter; the service sector, and specifically the disposal of waste in landfills, is the one that contributes most to

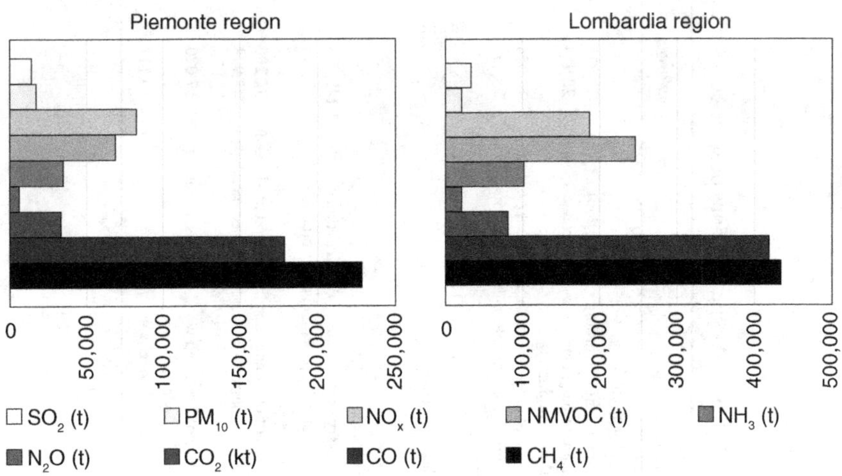

Figure 5.1 Pollutant emissions in Piemonte and Lombardia regions, 2005.

Table 5.2 NAMEA-type emissions of some economic activities from the secondary sector in Piemonte and Lombardia regions, 2005

	Employees	*Local units*	CH_4 *(t)*	*CO (t)*	CO_2 *(kt)*	N_2O *(t)*	NH_3 *(t)*	*NMVOC (t)*	NO_x *(t)*	PM_{10} *(t)*	SO_2 *(t)*
PIEMONTE REGION											
Total industry (not including construction)	**479,875**	**47,937**	**24,308.24**	**26,838.97**	**15,545.44**	**2,643.56**	**204.96**	**27,923.06**	**22,932.89**	**4,285.33**	**11,220.39**
Manufacture of pulp, paper and paper products; publishing and printing	22,984	2,634	80.47	4,318.64	844.27	24.72	49.08	6,472.06	1,995.78	1,295.49	167.29
Manufacture of other non-metallic mineral products	14,354	1,788	11.62	396.39	1,572.88	2,503.44	19.48	2,167.85	2,556.70	82.53	8,024.54
Manufacture of basic metals and fabricated metal products	93,457	11,569	31.34	497.50	2,897.92	20.48	1.42	340.41	6,455.22	404.80	2,032.76
Manufacture of wood and wood products, of rubber and plastic products	64,429	9,139	129.29	3,582.70	1,603.65	36.25	17.07	6,175.48	2,520.92	602.76	339.09
Electricity, gas and water supply	9,871	625	23,511.98	3,727.20	6,158.53	12.83	1.01	340.53	4,653.36	93.08	137.61
LOMBARDIA REGION											
Total industry (not including construction)	**1,149,448**	**122,684**	**106,278.74**	**49,798.69**	**36,258.35**	**1,186.53**	**354.07**	**98,843.37**	**56,844.84**	**3,839.36**	**24,092.89**
Manufacture of pulp, paper and paper products; publishing and printing	69,578	8,421	93.87	606.11	1,040.13	46.21	5.69	2,887.63	1,197.56	129.74	177.06
Manufacture of other non-metallic mineral products	31,206	3,628	82.69	1,267.18	4,214.88	60.41	81.02	1,074.46	4,176.37	410.68	5,212.33
Manufacture of basic metals and fabricated metal products	248,978	29,817	551.99	20,834.09	4,305.41	350.94	89.99	27,087.10	18,373.46	986.72	3,044.14
Manufacture of wood and wood products, of rubber and plastic products	157,020	23,166	127.31	4,603.94	927.84	77.17	93.29	23,950.28	3,518.54	352.40	157.09
Electricity, gas and water supply	19,111	1,016	104,567.56	3,669.48	19,996.34	202.17	13.40	4,850.89	14,663.95	538.83	11,757.47

Table 5.3 Impact of air emissions by macro-sectors in Turin and Milan provinces, 2005 (%)

	CH_4 (t)	CO (t)	CO_2 (kt)	N_2O (t)	NH_3 (t)	NMVOC (t)	NO_x (t)	PM_{10} (t)	SO_2 (t)	Greenhouse gases	Acidification	Tropospheric ozone layer
TURIN PROVINCE												
Agriculture	38.19	2.18	1.59	66.36	91.67	1.38	7.49	14.64	2.12	6.05	39.42	4.95
Industry	17.36	13.89	45.43	6.38	0.60	53.84	29.77	29.48	37.18	42.22	18.98	37.33
Services	42.74	13.25	8.61	10.91	1.48	20.37	23.61	12.87	8.64	11.65	14.40	21.53
Households	1.72	70.69	44.36	16.35	6.25	24.41	39.12	43.01	52.05	40.08	27.20	36.19
MILAN PROVINCE												
Agriculture	20.27	1.78	0.87	38.91	87.97	0.41	2.99	7.17	0.86	3.27	24.31	1.63
Industry	50.73	6.86	34.30	12.19	0.73	46.76	26.07	23.72	63.16	34.76	23.21	36.03
Services	26.77	15.89	18.52	16.94	1.53	7.97	41.34	26.62	6.64	19.03	27.93	21.57
Households	2.24	75.47	46.31	31.96	9.76	44.86	29.60	42.50	29.34	42.94	24.55	40.77

CH_4 emissions. Similarly, the shares of N_2O emissions ascribed to each sector vary substantially compared with that identified at regional level because of a non-uniform spatial location of source activities among the provinces. In the province of Turin, for example, households emit more SO_2 than the secondary sector.

In the province of Milan the agricultural sector is also no longer the main source of CH_4 and the provincial NAMEA table highlights that households and the tertiary sector are responsible for a much larger share of the total impact, compared with the secondary sector, than at regional level.

A horizontal comparison of the two provinces reveals that the main differences concern emissions of CO_2, NO_x and SO_2. Once again, the disaggregation of the secondary sector allows us to understand which activities are responsible for these differences.

Most CO_2 emissions in the province of Turin are due to the 'Production and distribution of electricity, gas, steam and water' activity. This activity appears more efficient in terms of CO_2 emissions in the province of Milan, where it is, however, the main emitter of SO_2.

Total emissions by the secondary sector in the two provinces are more or less equivalent (top part of Table 5.4). However, the total emissions generated by household consumption are, for most pollutants, substantially higher in the province of Milan – between 25 per cent (CO_2) and 160 per cent (NMVOC) higher – much more than justified by demographic variables alone (the province of Milan counts 3,170,000 inhabitants while the province of Turin has 2,295,000).

The above examples clearly show how critical information on a local scale may prove to be in order to correctly target pollution control policies. It helps decision-makers understand, for instance, whether they should act primarily on the production or on the consumption side, which sectors and productive compartments need to be targeted and at which jurisdictional level.

The choice, in this case study, of the provinces of Turin and Milan, the most important cities in Northern Italy in terms of economic activity and population density, hints at the importance of extending this exercise to a municipal level, particularly in view of the administrative, fiscal and management independence that metropolitan areas are in the process of acquiring in the decentralization and federalism process.

Inter-provincial environmental accounts: NAMEA-type air emissions accounts in Cuneo and Alessandria[12]

While the first case study concerned provinces driven by the secondary and tertiary sectors, in this section we consider two provinces (both in the Piemonte region) where the primary sector is the leading economic activity.

The first province we consider is Alessandria. It covers a surface area of 3,560 km^2 with a population of 413,000 inhabitants, and three-quarters of the territory is mountainous and hilly. The main economic activities are agriculture (herbaceous and woody crops) and mechanical, chemical and plastics processing. The second province we consider is Cuneo: the territory is mainly mountainous

Table 5.4 NAMEA-type emissions of some economic activities from the secondary sector in the provinces of Turin, Milan, Alessandria and Cuneo, 2005

	Employees	*Local units*	*CH_4 (t)*	*CO (t)*	*CO_2 (kt)*	*N_2O (t)*	*NH_3 (t)*	*NMVOC (t)*	*NO_x (t)*	*PM_{10} (t)*	*SO_2 (t)*
TURIN											
Manufacture of pulp, paper and paper products; publishing and printing	12,743	1,523	13.30	184.38	128.44	2.65	1.87	1,589.57	386.59	94.62	25.83
Manufacture of other non-metallic mineral products	4,426	698	6.51	92.49	72.89	2.42	0.49	95.63	571.20	113.82	56.63
Manufacture of basic metals and fabricated metal products	52,190	5,978	45.67	3,364.70	520.37	9.71	7.50	2,713.49	1,337.67	432.10	146.95
Manufacture of wood and wood products, of rubber and plastic products	26,505	3,759	36.60	1,762.00	309.28	9.81	22.67	2,286.64	881.90	610.80	58.47
Electricity, gas and water supply	5,899	230	12,234.53	2,128.16	4,212.10	6.47	0.37	216.49	3,274.11	52.65	1,158.09
MILAN											
Manufacture of pulp, paper and paper products; publishing and printing	40,402	5,122	50.97	156.10	306.23	12.21	0.99	1,622.90	316.44	20.37	30.22
Manufacture of other non-metallic mineral products	8,308	1,998	21.85	190.76	294.36	9.97	0.47	328.60	1,205.01	93.51	627.77
Manufacture of basic metals and fabricated metal products	65,588	1,715	81.28	674.22	330.92	29.27	2.77	8,084.49	1,095.91	98.02	625.18
Manufacture of wood and wood products, of rubber and plastic products	53,583	8,949	87.13	1,739.46	396.19	31.30	28.04	8,479.82	1,307.05	114.16	47.83
Electricity, gas and water supply	8,170	356	35,889.79	614.86	3,996.41	13.79	0.47	1,357.31	3,224.60	53.24	1,654.95
ALESSANDRIA											
Manufacture of pulp, paper and paper products; publishing and printing	1,314	212	1.13	33.72	26.47	0.34	0.29	384.74	27.69	4.78	3.96
Manufacture of other non-metallic mineral products	1,246	147	6.33	51.05	1,299.52	5.74	0.07	188.87	1,190.51	142.88	1,107.07
Manufacture of basic metals and fabricated metal products	6,961	866	2.09	311.84	32.96	0.58	0.28	346.09	279.82	24.27	22.09
Manufacture of wood and wood products, of rubber and plastic products	1,4572	2,040	5.79	55.11	87.08	1.50	0.28	855.39	122.51	184.84	23.83
Electricity, gas and water supply	856	92	2,453.65	324.49	364.24	0.73	0.07	31.55	267.90	6.80	7.28
CUNEO											
Manufacture of pulp, paper and paper products; publishing and printing	4,058	368	1.73	46.30	103.24	0.76	0.31	365.90	36.65	3.36	4.38
Manufacture of other non-metallic mineral products	4,725	416	8.12	48.76	1,320.83	7.02	0.08	92.23	3,942.92	95.24	703.84
Manufacture of basic metals and fabricated metal products	9,929	1,595	8.05	1,335.79	287.53	3.87	0.29	300.49	330.60	273.47	29.99
Manufacture of wood and wood products, of rubber and plastic products	12,155	1,343	4.46	63.13	116.27	1.99	0.30	2,462.88	107.70	192.40	12.49
Electricity, gas and water supply	1,114	113	2,520.16	426.10	531.98	0.96	0.07	25.38	250.54	7.40	2.20

and covers a surface area of 6,903 km^2 with a population of roughly 583,000 inhabitants (almost double the territorial dimension of the province of Alessandria, with about half the population density). The leading economic sector is agriculture (crops and pastures) followed by confectionery and dairy products.

Comparing the percentage polluting impact of economic macro-sectors in the two provinces (bottom part of Table 5.4) highlights that most of the CH_4 emissions are due to agriculture in the province of Cuneo and to the service sector in the case of Alessandria. Within the service sector, this is due mainly to the presence of landfills and energy production. In addition, the module reveals that the percentage impact of polluting emissions in the case of Cuneo is much larger than the percentage impact at the regional level, not only for CH_4 but also for N_2O, PM_{10} and NH_3, pollutants mainly due to the use of chemical fertilizers (Table 5.5). Figure 5.2 shows the amount of pollutants in absolute terms.

In the case of Alessandria province, the analysis shows the increase in emissions ascribable to the higher population density: more urban waste directed to

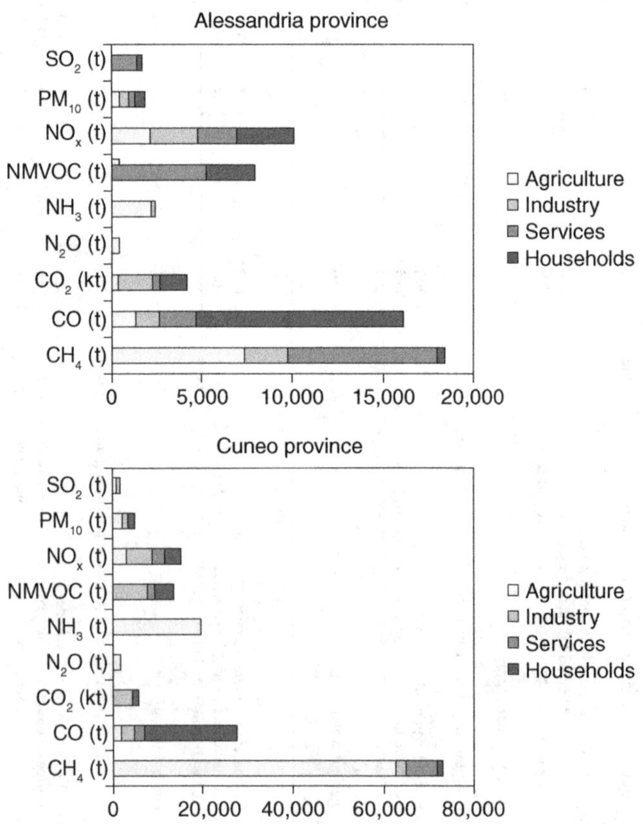

Figure 5.2 Pollutant emissions by macro-sectors in Alessandria and Cuneo provinces, 2005.

Table 5.5 Impact of air emissions by macro-sectors in Piemonte region and Alessandria and Cuneo provinces, 2005 (%)

	CH_4 (t)	*CO (t)*	*CO_2 (kt)*	*N_2O (t)*	*NH_3 (t)*	*NMVOC (t)*	*NO_x (t)*	*PM_{10} (t)*	*SO_2 (t)*	*Greenhouse gases*	*Acidification*	*Tropospheric ozone*
PIEMONTE REGION												
Agriculture	63.3	3.8	2.7	44.4	95.0	2.8	13.3	22.3	1.2	11.5	51.7	9.4
Industry	10.6	17.0	47.9	45.8	0.6	43.4	30.6	27.1	78.7	43.5	21.0	33.5
Services	24.8	10.1	7.3	3.4	1.2	15.5	20.2	11.4	3.1	9.1	9.2	17.6
Households	1.3	69.1	42.1	6.4	3.1	38.3	35.8	39.2	17.1	35.8	18.0	39.5
ALESSANDRIA PROVINCE												
Agriculture	39.4	7.6	4.1	82.8	91.1	4.7	20.9	19.0	1.9	9.1	41.9	14.3
Industry	13.4	8.5	50.6	3.6	0.2	46.0	26.7	26.4	73.9	46.3	23.9	32.0
Services	45.6	12.7	6.9	3.9	3.4	14.7	20.6	13.7	3.1	10.0	12.6	18.2
Households	1.6	71.2	38.4	9.6	5.3	34.6	31.8	40.9	21.2	34.6	21.6	35.6
CUNEO PROVINCE												
Agriculture	85.3	6.2	4.0	89.4	98.8	4.2	19.2	37.6	3.3	23.4	78.7	14.3
Industry	3.6	9.7	56.3	3.1	0.1	51.2	38.5	26.3	66.6	43.9	10.4	39.8
Services	0.03	4.3	3.7	0.7	0.04	10.7	11.3	6.1	3.4	2.8	2.6	10.1
Households	0.7	75.1	34.1	5.8	0.7	32.0	26.8	28.6	25.6	26.4	7.1	32.1

landfills (CH_4) and higher impact of transport (CO, CO_2, NMVOC, NO_x) when compared to Cuneo. A more detailed analysis of the impact from households (Figure 5.3) allows us to attribute around 30 per cent of NMVOC, NO_x and PM_{10} emissions and 50 per cent of CO emissions to the transport sector. Transport and heating account for 70 per cent of emissions of these pollutants.

In the province of Cuneo, although the total ratio of CO emissions is almost equal to that of Alessandria, we can see how heating weighs more than proportionally if we consider the total population living in this province. This can be explained by the residential sprawl that characterizes the area: scattered housing requires more dispersive heating systems.

In Table 5.4, the NAMEA for the province of Alessandria disaggregated for the secondary sector shows high emissions of CO_2 due to the use of combustion engines for the processing of cement; combustion processes using fossil fuels generate high emissions of SO_2.

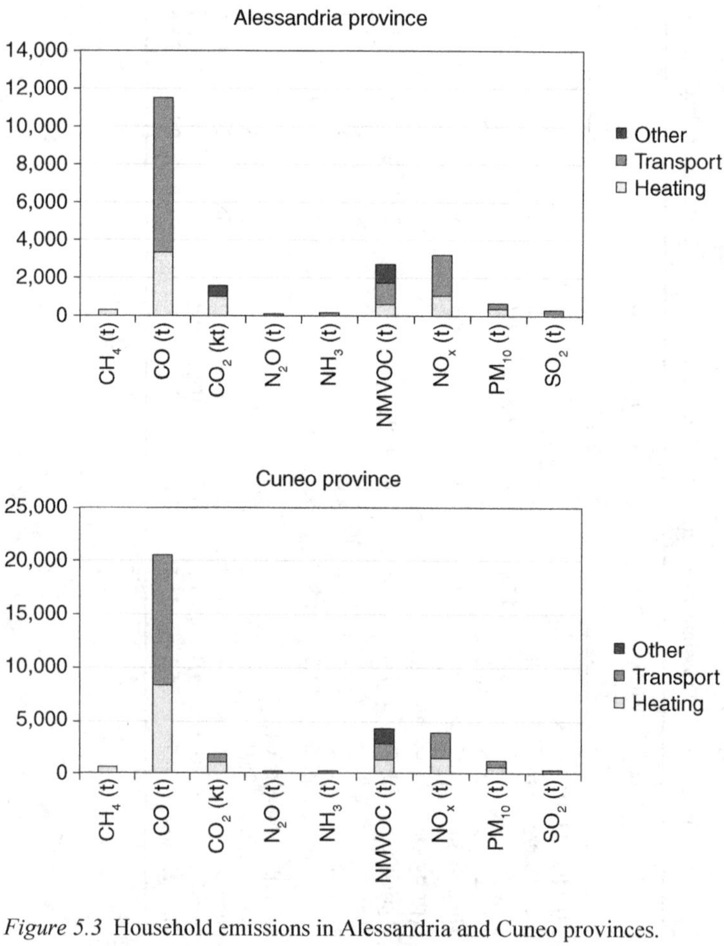

Figure 5.3 Household emissions in Alessandria and Cuneo provinces.

In Table 5.4, the NAMEA for the province of Cuneo disaggregated for the secondary sector shows high emissions of CO_2 and SO_2 generated by 'Manufacture of other non-metallic mineral processing', while the main sector in terms of NMVOC emissions is 'Manufacture of wood, rubber, plastics and other manufacturing'.

A horizontal comparison within the Piemonte region of the three considered provinces (Turin, Alessandria and Cuneo) clearly highlights very different environmental issues and, equally importantly, a very different sectoral allocation of responsibility. Again, uniform policies designed on a regional scale on the basis of aggregate information can only be inefficient.

Intra-provincial and inter-municipal environmental accounts: NAMEA-type air emissions accounts in the municipalities of Robilante and Morozzo in the province of Cuneo[13]

Let us now consider the municipal level. We present the case of two small municipalities within the province of Cuneo, similar in terms of surface area, about 20 km^2, and number of inhabitants, about 2,000, but very different in terms of economic activity.

The development of Robilante has been driven by the industrial sector: a ceramic and a cement factory, processing the raw materials which come from an adjacent municipality. In addition it hosts a ski resort. The economy of the town of Morozzo, on the other hand, is based mainly on breeding (capons) and on the presence of a natural reserve. The two municipalities are thus very different from an economic, social and environmental point of view. The impact of emissions disaggregated by macro-sector for the municipalities of Robilante and Morozzo is shown in Table 5.6.

The different productive structure is reflected in very different patterns of environmental impact: the municipality of Robilante records CO and CO_2 emissions which are ten times higher than those of Morozzo, and NO_x emissions which are 70 times those of Morozzo. On the other hand, Morozzo records high emissions of CH_4 and NH_3.

A more detailed analysis, disaggregating the secondary sector, confirms that most of CO, CO_2 and NO_x emissions in the jurisdiction of Robilante are generated by the compartment to which the ceramic and cement factories belong. The indicators reported in Table 5.6 turn out to be closely linked to 'Manufacture of other non-metallic mineral processing' that emits 98 per cent of greenhouse gases (CO_2) and contributes to acidification and formation of trophospheric ozone with emissions of NO_x and CO.

This third case study draws attention to the existence of local contexts, even within provinces, that may require targeted polices. Although the province of Cuneo is considered mainly agricultural when compared with other provinces such as Turin or Alessandria, there may still be municipalities within it dominated by a manufacturing sector which is the source of serious local impacts, as demonstrated in the case of Robilante. These contexts will require different environmental management and policies compared with the surrounding areas.

Table 5.6 Air emissions by macro-sectors in the municipalities of Robilante and Morozzo

	CH_4 (t)	*CO (t)*	*CO_2 (kt)*	*N_2O (t)*	*NH_3 (t)*	*NMVOC (t)*	*NO_x (t)*	*PM_{10} (t)*	*SO^2 (t)*	*Greenhouse gases*	*Acidification*	*Tropospheric ozone*
ROBILANTE												
Agriculture	78.41	5.79	0.22	2.99	17.66	1.24	2.71	0.34	0.04	2,792.83	35.49	6.28
Industry	0.09	1,686.87	1,256.22	0.03	0.02	5.26	4,948.10	89.59	7.42	1,256,232.59	3,471.12	6,227.49
Services	0.06	7.75	0.84	0.05	0.03	1.96	4.73	1.09	0.17	854.66	3.54	8.58
Households	2.11	110.76	21.41	0.39	0.31	20.56	16.11	6.15	1.26	21,575.74	13.13	52.42
MOROZZO												
Agriculture	624.76	22.81	3.42	14.96	227.94	8.07	44.64	6.42	0.61	21,176.82	464.95	73.79
Industry	0.06	4.00	0.88	0.03	0.01	5.65	2.91	0.55	0.25	889.00	2.32	9.65
Services	0.35	6.29	0.82	0.06	0.02	7.67	6.57	1.26	0.17	844.92	4.81	16.39
Households	1.34	72.78	6.10	0.30	0.20	14.42	11.98	4.25	1.14	6,221.78	9.90	37.07

Further analyses enabled by local NAMEA-type accounts

Shift-share analysis

The case studies presented in the previous section are based on very simple descriptive analyses that nonetheless may already represent a powerful information tool. More structural analyses are possible, even though at present the limiting factor of NAMEA-type analytic applications at sub-regional level is the lack of time series. Not all the analyses currently undertaken at national level in EU member states can be conducted on a sub-regional scale, but some can, and, in our opinion, with useful results. Here we propose the application of a vertical shift-share analysis for the Piemonte region, the province of Turin and the municipality of Turin[14] for the year 2005.

Shift-share analysis has already been undertaken employing NAMEA outcomes (Mazzanti *et al.*, 2007; Dosi *et al.*, 2008; van Rossum and van de Grift, 2009; Bonazzi, 2009). The main policy issues that this kind of technique is supposed to address are those related to the analysis of economic growth/decline of an area, the state of a community compared to other communities, the economic sectors to be monitored or subsidized, and so on. Through shift-share analysis, the role of economic activities can be isolated and the gap between emission efficiency in the different sectors explained at different administrative levels. In this application, we compare the regional, provincial and municipal levels (instead of national and regional levels as is usually done) and we use the number of employees instead of value added in the calculation of emission intensity.[15]

Following the application proposed by Esteban (2000) we calculate three indicators: (a) the industry mix, which describes how specialized the economic system is in some economic activities; when negative, this indicator indicates that, at a sub-hierarchical level, the sectors that employ more workers are less polluting; (b) the productivity differential, which compares the efficiency of a sub-hierarchical level with the superior one; when negative, this indicator indicates that, at sub-hierarchical level, economic activities pollute less than at the higher hierarchical level; (c) the allocative component, which presents the contribution of sub-hierarchical levels to economic activity where the higher one shows a higher performance; when negative, this indicator indicates that the sub-hierarchical level is specialized in the economic activities that pollute less.

The composition of economic sectors in terms of workers employed varies substantially across the three administrative levels. In some activities the number of employees grows in relative terms from the regional to the municipal level (real estate, information technology, etc.), in other activities it decreases (manufacture of machinery) and in others it remains almost unchanged (e.g. wholesale and retail).

Sectoral emissions at the regional and provincial levels are related to the number of employees in Table 5.7. Here the analysis highlights how, in terms of emissions of CH_4 and CO per unit of labour, the economic activities in the province of Turin are less efficient than those in the Piemonte region as a whole.

Table 5.7 Shift-share coefficients for the economic system region–province, 2005

	Xreg	*Xprov*	*Xp–Xr*	*im*	*pd*	*Ac*	*im + dp + ac*
CH_4	73.35	85.62	12.28	4.80	7.33	0.15	12.28
CO	24.98	26.30	1.32	1.32	0.07	–0.07	1.32
CO_2	11.79	9.46	–2.33	0.13	–3.32	0.86	–2.33
N_2O	14.01	0.16	–13.86	–5.21	–13.87	5.22	–13.86
NH_3	0.37	0.17	–0.20	–0.04	–0.18	0.03	–0.20
NMVOC	27.52	23.50	–4.02	–2.23	–3.32	1.52	–4.02
NO_x	27.58	20.79	–6.78	–1.08	–8.16	2.46	–6.78
PM_{10}	5.93	4.71	–1.22	–0.29	–1.05	0.13	–1.22
SO_2	8.77	2.00	–6.77	–2.23	–6.93	2.39	–6.77

By comparing region–municipality (Table 5.8) and province–municipality (Table 5.9), the three indicators emerge as negative only for SO_2. For all other pollutants the sub-hierarchical level does not prove to be more efficient when compared to the superior hierarchical level. This is probably driven by the choice of Turin as case-study province and municipality.

In the case of CH_4 and CO neither the sector specialization nor pollutant emissions are ever efficient – we have a positive sign for the indicators 'industry mix' and 'productivity differential' in almost all the administrative levels considered (Table 5.7, Table 5.8 and Table 5.9). The economic activities in the municipality of Turin are not specialized in sectors with the highest environmental efficiency for CH_4, CO and NH_3 and are less efficient than the regional and provincial levels for the emissions of CH_4, CO and NMVOC. Local decision-makers could act in several different directions: from promoting eco-efficiency techniques in the activities that play a leading economic role at municipal level to supporting the development of sectors that prove to be more environmentally friendly.

Linking emissions and concentrations through chain modelling

The compilation of NAMEA-type accounts entails working with local technical and policy units since this enables the analyst not only to understand the real

Table 5.8 Shift-share coefficients for the economic system region–municipality, 2005

	Xreg	*Xcom*	*Xc–Xr*	*im*	*pd*	*Ac*	*im + pd + ac*
CH_4	73.35	118.60	45.26	17.55	22.53	5.18	45.26
CO	24.98	30.10	5.12	1.52	9.17	–5.58	5.12
CO_2	11.79	4.35	–7.44	–2.41	–6.71	1.68	–7.44
N_2O	14.01	0.05	–13.96	–6.78	–13.98	6.79	–13.96
NH_3	0.37	0.02	–0.35	0.02	–0.34	–0.03	–0.35
NMVOC	27.52	22.57	–4.96	–7.99	6.40	–3.36	–4.96
NO_x	27.58	20.10	–7.48	–2.54	–4.62	–0.31	–7.48
PM_{10}	5.93	4.28	–1.65	–0.88	–0.79	0.02	–1.65
SO_2	8.77	1.39	–7.38	–3.72	–7.00	3.34	–7.38

Table 5.9 Shift-share coefficients for the economic system province–municipality, 2005

	Xprov	*Xcom*	*Xc−Xp*	*im*	*pd*	*Ac*	*im+pd+ac*
CH_4	85.62	118.60	32.98	16.82	16.56	−0.40	32.98
CO	26.30	30.10	3.80	−0.57	8.81	−4.44	3.80
CO_2	9.46	4.35	−5.11	−2.21	−3.78	0.88	−5.11
N_2O	0.16	0.05	−0.11	0.02	−0.12	−0.01	−0.11
NH_3	0.17	0.02	−0.15	0.01	−0.15	−0.02	−0.15
NMVOC	23.50	22.57	−0.93	−5.64	10.80	−6.10	−0.93
NO_x	20.79	20.10	−0.69	−0.86	3.80	−3.63	−0.69
PM_{10}	4.71	4.28	−0.43	−0.53	0.51	−0.41	−0.43
SO_2	2.00	1.39	−0.61	−0.52	−0.02	−0.08	−0.61

demands of policy-makers and, based on those, to focus on data that provide an answer, but also to combine the proposed environmental accounting tools with local knowledge and expertise to develop practically implementable integrated tools.

Local authorities are likely to aim at acting on specific air pollutants. In this case, policy-makers should consider both emission sources and pollutant concentrations. The EU Air Quality Framework and Directives make it compulsory for member states to assess air quality within their territories. Air Quality Assessment is undertaken through local observation of pollutant concentrations at sampling sites, but jurisdictions are not fully covered by the sampling network. Local authorities therefore use air-quality models whose reliability and accuracy is verified by comparing modelling results with real measurements. In Piemonte, the regional environmental protection agency (Arpa) has developed a 3-D modelling system that simulates air pollution emissions, transport, diffusion and chemical reactions, and concentrations for the pollutants CO, NO_x, SO_2, PM_{10}, $PM_{2.5}$, O_3 and benzene (C_6H_6). The simulation is based on a deterministic modelling system which consists of an emission processing system (named EMMA; see Arianet, 2005), a diagnostic meteorological model (named MINERVE; see Aria Tech. 2001), an atmospheric turbulence and dispersion parameter interface module (named GAP/SurfPRO; see Finardi *et al.*, 2005) and a Eulerian chemical transport model (named FARM; see Cost728, 2006).[16] The territorial units employed are municipalities and the results of simulation refer to grid cells. The common ground that allows us to compare the NAMEA-type outputs with the chain modelling results is the IREA database which constitutes the data source for both. Given the complexity of the processes that lead to pollutant concentrations, the relationship we can identify between emissions and concentrations is non-linear and the measurement units will not be the same. Maps showing where the most polluting source is located and where the highest concentrations are recorded can, however, provide local policy-makers with some input for planning. They enable, for example, seeing where the target does not have to be attained for each pollutant, which administrative level has to be involved and, especially at the borders between regions, where inter-regional cooperation should be sought.

Figures 5.4–5.7 allow us to visually compare emissions based on IREA and concentrations resulting from chain modeling for four pollutants (in 2005).[17]

The impacts of NO_2 (Figure 5.4) and NO_x (Figure 5.5) affect large areas around the emission source, and emission sources in one municipality may affect many provinces. Pollution control policies in one municipality, therefore, will have beneficial effects spilling over several other jurisdictions.

In addition, critical pollutant concentrations may be located in areas where there is no relevant emission source, as shown in the case of PM_{10} (Figure 5.6) and SO_2 (Figure 5.7). Especially where more than one region is involved, citizens' health protection will require inter-regional cooperation.

Information on the drivers of polluting emissions made available to local policy-makers would allow them to assess the relative merits of alternative policy options properly (promotion of eco-efficiency practices in production activities, enhancement of infrastructures and transport facilities, choice of sectors to be supported with incentives, and so on) and implement better targeted, cost-efficient and effective actions for protecting environmental quality.

Final remarks

The range of information that can be obtained from the application of integrated environmental and economic accounts of the NAMEA type is large, and it is

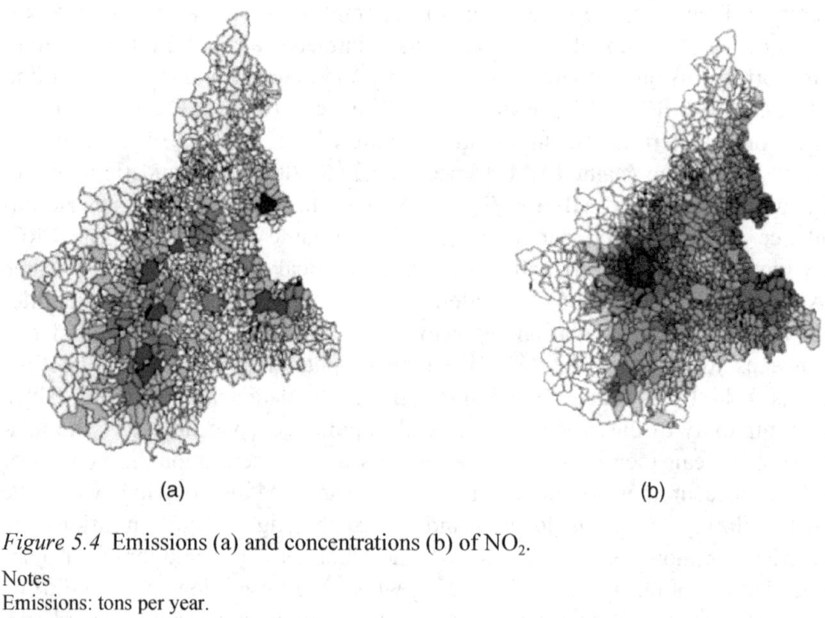

Figure 5.4 Emissions (a) and concentrations (b) of NO_2.

Notes
Emissions: tons per year.
Concentrations: yearly average of μg/mc.

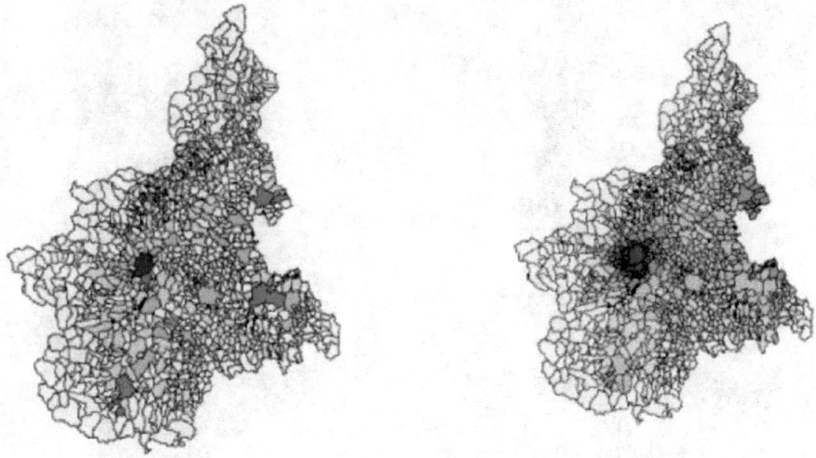

(a) (b)

Figure 5.5 Emissions (a) and concentrations (b) of NO_x.

Notes
Emissions: tons per year.
Concentrations: yearly average of µg/mc.

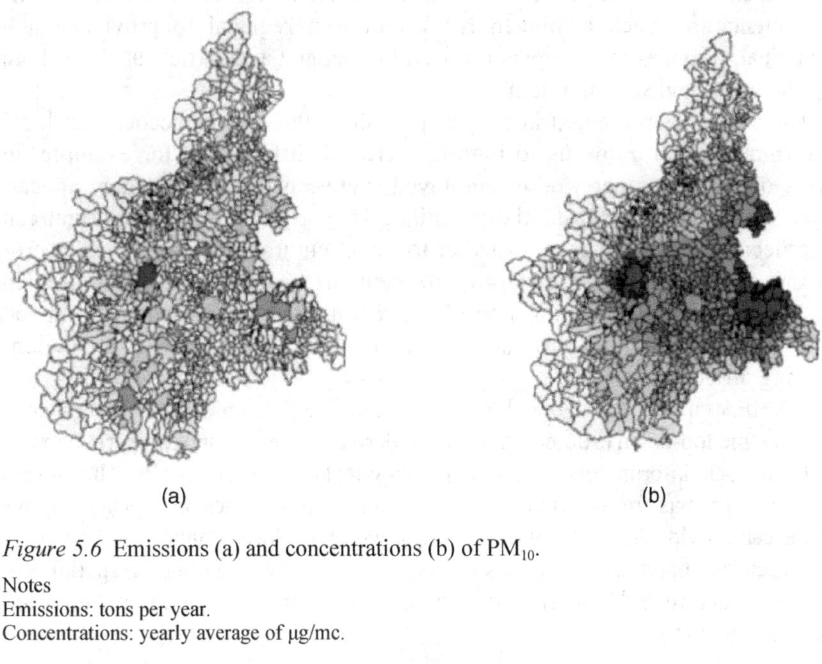

(a) (b)

Figure 5.6 Emissions (a) and concentrations (b) of PM_{10}.

Notes
Emissions: tons per year.
Concentrations: yearly average of µg/mc.

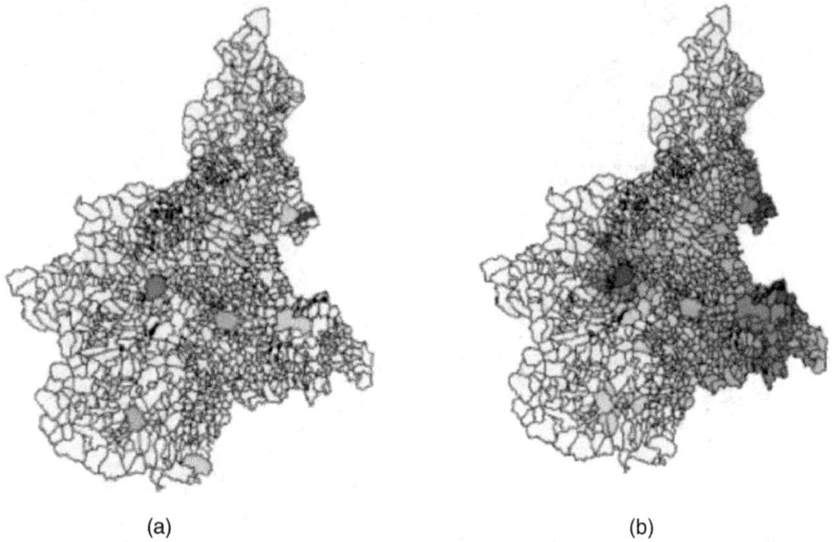

Figure 5.7 Emissions (a) and concentrations (b) of SO_2.

Notes
Emissions: tons per year.
Concentrations: yearly average of μg/mc.

difficult to establish a priori which local administrative level is ideal for its implementation. Each administrative level, from regional to provincial and municipal, provides many inputs for useful horizontal and vertical analyses both in descriptive and structural terms.

The examples presented in this chapter show that hybrid accounts at local government level allow us to highlight crucial differences (for example, in terms of emissions per worker employed) between regions that may appear, from aggregate data, substantially similar. They provide information that can help decision-makers choose whether to direct environmental policies to producers or households. They draw attention to the territorial differences in terms of prevailing economic activities and consequent environmental impact, both among and within provinces. They identify spillovers between neighbouring jurisdictions.

NAMEA compiled at national level is obviously a precious tool for a number of uses, but the top-down estimates that can be derived from it cannot properly approximate the local information that can help policy-makers who act locally. The various territorial contexts are so different that only bottom-up analyses on a local scale, we argue, can provide detailed and reliable information for local planners.

Structural analytical techniques can be useful for investigating the spatial production structure and help effective policies for local development and environmental protection.

The validity of the assessment tools provided by environmental accounting can be further strengthened by combining them with knowledge and expertise of the technical units working in closer contact with the territory, in local government environment departments and environmental protection agencies.

There are still many shortcomings in the application of NAMEA-type accounts at local level: database incompleteness (ASIA does not provide value added and production data), lack of time series (IREA is only available for a few years so far and ASIA has only been available since 2004), non-automation of the compilation process (mainly because in some phases local technical units must provide information in order to make meaningful assignments of emissions to sources). The advantages in terms of informational content of the output, however, turn out to be potentially extremely interesting for the design, management, implementation and assessment of environmental policies at local level, since the subsidiary approach applied to environmental policies at the EU level assigns the large part of competences and responsibilities to local authorities.

Notes

1 See www.istat.it/dati/dataset/20090401_00.
2 All data and accounting modules that have been omitted from the chapter due to space contraints are available from the authors on request.
3 Source of data should specify whether the responsibility of providing the information lies with a public institution, a voluntary based initiative, a private organization and so on and the purpose for which data are collected.
4 Processing procedures describe the way data are collected, classified, processed and stored and they have to be transparent.
5 Aggregating small consistent units in order to obtain a larger administrative/territorial unit is possible and advisable, but the disaggregation from a larger scale aimed at estimating values for smaller units exposes results to several criticisms.
6 Sources must be able to provide up-to-date data and to maintain a regular, adequate periodicity.
7 It includes 11 macro-sectors, 75 sectors and 430 activities and it is available for the years 2001, 2005 and 2007. The pollutants recorded are methane (CH_4), carbon monoxide (CO), carbon dioxide (CO_2), nitrogen dioxide (N_2O), ammonia (NH_3), non-methane volatile organic compounds (NMVOC), oxides of nitrogen (NO_x), sulphur dioxide (SO_2) and particulate matters (PM_{10}).
8 For more details on the methodology, see Dalmazzone and La Notte (2009).
9 The formula for calculation of Global Warming Potential is $CO_2*1000+310*N_2O+CH_4*21$.
10 The formula for calculation of acidification is $SO_2+0.7*NO_x+1.9*NH_3$.
11 The formula for calculation of tropospheric ozone formation is $NMVOC+1.22NO_x+0.11*CO+0.014*CH_4$.
12 Data collected and partially processed by Diano, Locatelli and Roatta, University of Turin.
13 Data collected and partially processed by Belgero, Fregnan and Nada, University of Turin.
14 The following is an extract from a previously published case study (La Notte *et al.*, 2009).
15 We cannot use value added because data are not available in the ASIA dataset. Previous

studies (e.g. Harrison and Antweiler, 2003; Hettige *et al.*, 1995) have used the number of employees as an acceptable proxy for similar calculations.

16 For further details, see Arduino *et al.* (2007) and La Notte *et al.* (2009).

17 The software used is ArcGIS 9.3 and for the visual representation we adopt the natural break classification criteria in five classes.

References

Arduino, G., Contardi, C., Bovo, S., Bande, S., Clemente, M., Calori, G., De Maria, R., Finardi, S., Muraro, M. and Silibello, C. (2007) 'Utilizzo dei GIS nell'ambito della Valutazione sulla Qualità dell'Aria', paper presented at Atti 11° Conferenza Nazionale ASITA, Centro Congressi Lingotto, Turin, 6–9 November.

ARIANET (2005) *EMMA (EMGR/Make) User's Guide Version 3.5*, Rapporto Arianet R2005.08, Milan.

ARIA Technologies (2001) *MINERVE Wind Field Model General Design Manual*, ARIA Tech. Report, Paris.

Bonazzi, E. (2009) 'Emissioni di gas serra ed eco-efficienza: il vantaggio dell'Emilia-Romagna', *Arpa Rivista*, 1: 48–49.

COST728 (2006) 'Cost 728/732 model inventory'. Online: www.mi.unihamburg.de/cost-modin.

Dalmazzone, S. and La Notte, A. (2009) 'Environmental accounting at different levels of government: the state of the art', in A. Breton, G. Brosio, S. Dalmazzone and G. Garrone (eds), *Governing the Environment: Salient Institutional Issues*, Cheltenham: Edward Elgar.

Dosi, M.P., Bonazzi, E. and Sansoni, M. (2008) 'Progettare la sostenibilità nello sviluppo di un territorio: l'analisi Shift-Share su aggregati economico-ambientali', paper presented at the XXIX Conferenza Italiana di Scienze Regionali, Bari, September.

Esteban, J. (2000) 'Regional convergence in Europe and the industry mix: a shift-share analysis', *Regional Science and Urban Economics*, 3: 353–364.

European Commission (2009) *Manual for Air Emissions Accounts*, KA-RA-09-004-EN, Brussels: European Commission.

Faramondi, A. (2008) 'Valore aggiunto comunale: integrazione tra fonti e approccio bottom-up', *Rivista Internazionale di Scienze Sociali*, 116: 179–209.

Finardi, S., De Maria, R., D'Allura, A., Cascone, C., Calori, G. and Lollobrigida, F. (2005) 'A deterministic air quality forecasting system for Torino urban area', *Environmental Modelling and Software*, 23: 344–355.

Harrison, K. and Antweiler, W. (2003) 'Incentives for pollution abatement: regulation, regulatory threats, and non-governmental pressures', *Journal of Policy Analysis and Management*, 3: 361–382.

Hettige, H., Martin, P., Singh, M. and Wheeler, D. (1995) 'IPPS: The Industrial Pollution Projection System,' World Bank, Policy Research Department Working Paper, February, Washington, DC: World Bank.

La Notte, A. (2008) 'Integrated environmental and economic accounts applied at local level: Results of NAMEA flow accounts at sub-regional level in Piemonte (Italy) – air emission origin and destination through an atmospheric modeling system', Proceedings of the 10th Biennial Conference of the International Society for Ecological Economics, Nairobi, Kenya, 7–11 August.

La Notte, A., Arduino, G., Sordi, F. and Truffo, G. (2009) 'La contabilizzazione di emissioni e concentrazioni degli inquinanti in atmosfera utilizzando conti ambientali e chain modeling', *Economia delle Fonti di Energia e dell'Ambiente*, 3.

Mazzanti, M., Montini, A. and Zoboli, R. (2007) 'Struttura produttiva territoriale ed indicatori di efficienza ambientale: un'analisi attraverso la NAMEA regionale', *Economia delle Fonti di Energia e dell'Ambiente*, 1: 35–64.

United Nations (UN) (2003) *SEEA: System of Integrated Environmental and Economic Accounting (an Operational Manual)*, New York: United Nations Publications.

Van Rossum, M. and van de Grift, M. (2009) 'Regional analysis: differences in emission-intensity due to differences in economic structure or environmental efficiency?' *Journal of Sustainable Development*, 2: 43–56.

6 Air emissions and displacement of production

A case study for Italy, 1995–2007

Renato Marra Campanale and Aldo Femia

Introduction

In a previous study by the same authors, the general factors behind the trends of Italian production-related air emissions relevant to three environmental themes, greenhouse gases (GHG), acidifying substances and tropospheric ozone precursors, were analysed for the period 1992–2006 by means of decomposition analysis (Femia and Marra Campanale, 2010). Hybrid environmental and economic accounts (National Accounting Matrix integrated with Environmental Accounts, NAMEA-type tables), including data on energy use by activity, were used to break down the yearly emission variations into changes in four determinants: level of economic activity, structure of production, energy use per output unit (energy intensity of output) and emissions per unit of energy input (emission intensity of energy use).

For all three environmental themes, results showed that total emission changes had been mainly held low by virtue of the two components representing the environmental efficiency of industries. With regard to the structure of production, it had a generally favourable impact on total emission variation.

Two questions arise from this analysis. Is a reduction in the emission intensity of energy use mainly due to a better use of the same fuels or has a shift towards less polluting fuels taken place? Is the reduction in the energy intensity of output a consequence of technological and organizational improvements or is it due to the fact that more energy-intensive goods are imported rather than produced by domestic industries?

In this chapter, we address the second question. It is a well-known fact that more and more of Italian industry products are actually being manufactured abroad, with only the final stages being performed in Italy; this implies that the output is obtained without using the necessary energy in the 'producing' industry which in reality only buys, as intermediate inputs, products that are almost finished from abroad and resells them after little transformation. This displacement of production may be an important explanatory factor for the observed environmental Kuznets curve (EKC) for air emissions; its importance depends directly on the pollution-intensiveness of the displaced industries. To the extent that this explanation holds, the EKC pattern does not deliver very significant information on the environmental effects of economic growth as regards the global dimension. In this chapter, we explore this issue with reference to the most global of environmental issues, climate change. We do this by quantifying and comparing: (a) the total GHG emissions that would have been generated by Italian producers if all products needed to satisfy the final demand of Italians as well as foreign demand for Italian products had been produced in Italy; (b) the changes in these hypothetical emissions that can be ascribed to genuine technological changes and to changes in the level and composition by industry of the various types of final demand (private and public consumption, gross capital formation and exports) respectively; (c) the parts of these hypothetical emissions ascribable to intermediate and final imports respectively; (d) the effects of the shift between the sources, whether domestic or foreign, of final and intermediate products on the import-driven parts of the hypothetical emissions.

We apply a structural decomposition analysis (SDA) to an environmentally extended input–output (EE-IO) model based on NAMEA[1] and on supply-and-use tables made publicly available by the Italian National Institute of Statistics (ISTAT, 2010, 2011), supplemented by additional information on some industries that are not significantly present in Italy.

Methodological framework

Perspectives for addressing environmental pressures and policy consequences

Environmental pressures can be analysed following two approaches based on the Input–Output Framework of the European System of Accounts (Eurostat, 2008). In the 'responsibility of the producer' or 'direct flow' approach, the pressures are those of 'standard' industries, i.e. industries defined as the sets of all statistical units that carry out the same (main) economic activity. The statistical data on the environmental pressures by industry made available by European National Statistical Offices (ISTAT for Italy) through the NAMEA framework respond to this straightforward approach. On the other hand, according to the 'responsibility of the final user' or 'total flow' approach, the environmental pressures refer to the 'vertically integrated' industries (Pasinetti, 1973), i.e. to the sets of all production activities that are directly and indirectly needed to obtain final products. In this case, for each industry, the focus is only on the whole production chain of its final products. This is a broad process involving the entire production system. Therefore, although a vertically integrated industry takes its name from its final products, it is in reality very different from the 'standard' industry of the same name since it is a collection of extremely diverse activities. Each vertically integrated activity is completely independent of the others and contains all that is needed from extraction to sale to the final user (Femia and Panfili, 2005). Its emissions are calculated by cumulating the emissions of all the parts of the 'standard industries' that contribute to the final result, from the extraction of the necessary natural resources up to the delivery of the final product. Indeed, data responding to the 'total flow' approach may only be a mathematical artefact, i.e. they are not observable data but the result of a calculation based on a NAMEA-like description of the direct pressures which is therefore a prerequisite for the calculation of 'total flows' through the environmentally extended Leontievian model, along with the description of the inter-industry structure provided by the supply-and-use and input–output tables. This study takes the latter approach.

The choice between the two perspectives mainly depends on the driving force, production or final use, one wishes to take as an explicit policy target. Under the 'direct flow' approach, the level and composition of the final uses is not the possible aim of policies, whereas the environmental efficiency of individual (standard) industries is at the centre of attention. On the contrary, under the 'total flow' approach, the question is how the level and composition of the final uses should change if we want to achieve a given change in environmental pressures.

This choice is sometimes seen as a judgement on who is responsible for environmental pressures, whether it is the producers or the users of the final products. Clearly, the purchase of a final good activates the pressures of the whole production chain so that the choice of whether to buy a final product may be laden with an ethical value with respect to the emissions caused. However, this is not a necessary feature of the approach, at least as long as the total environmental pressures stemming from national production are merely reclassified (the total amount being the same at the domestic *macro*, economy-wide, level) according to their functional value. However, the 'total flow' approach allows us to go beyond domestic boundaries since the application of its logic can be easily extended to the production of traded products in order to focus on *all* the environmental pressures caused by final purchases, including the pressures directly stemming from foreign production systems. In this case, the 'vertically integrated' industries will constitute the pressures 'embodied' in the imported products or the pressures avoided. The study of the relationship between the direct pressures of the Italian economy and those avoided due to trade is the specific aim of this study.

In recent years, this question has been posed in connection with global climate negotiations.[2] In this case, the focus is on GHG since for these pressures on the global environment, the possibility of shifting their generation to other countries by relying on international trade and outsourcing may weaken the environmental justice of the agreements based on direct responsibility only (Peters and Hertwich, 2008). The emphasis on air emission measures responding to the direct flow perspective (i.e. only measuring domestic emissions) might induce the countries who should reduce their emissions to offshore/outsource their production to countries where GHG targets do not apply. This suits industrialized countries whose final and intermediate demand heavily drives the production-related emissions in export-oriented economies such as China.[3]

The international displacement of industrial production calls attention to the shift of high polluting industries to countries with lower environmental standards. This elusive phenomenon by which developed countries achieve their environmental targets more easily than they would otherwise may even result in a negative overall effect. Therefore, addressing the issue of 'responsibility' means considering the fact that some of the commodities produced by a country are consumed abroad as well as the fact that some of its final use of products depends on production, and hence emissions, abroad, and accounting for the emissions concerns the whole production chain, no matter where goods and services are produced.

Offshoring and displacement

The internationalization of economic activities is an essential feature of globalization of the economy. As Costa and Ferri (2007) write, 'several theoretical and empirical studies on the domestic effects of offshoring of production follow the political and social concern that relocating part of business abroad depletes employment and worsens performance at home'. This phenomenon is usually measured by taking account of the share of imports in the intermediate inputs of the domestic production process.

This chapter deals with the importance of delocalization as a possible driver of improvements in Italian energy efficiency. We investigate to what extent efficiency improvement is explained by the displacement of Italian industrial production, made operational as a composition effect, as an increase in the share of imports in intermediate and final demand. Then, the notion of displacement we adopt captures the shift abroad of energy-intensive industries and stages of production which improves the composition by industry as well as the energy efficiency of domestic activities, although it does not improve the structure of final demand.

EE-IO analysis of the total GHG emissions 'embodied' in Italian industries' final products and in imports

The model

The EE-IO model calculates the total GHG emissions associated with Italian economic activities, including those avoided due to international trade, by analysing the direct and indirect emissions required to produce the final products of Italian industries (including those activated by final demand through imported intermediate inputs) as well as the products imported for final uses.[4]

For the purposes of this calculation, we integrate the Italian IO tables ad hoc in order to supplement the information on the cost structure of some industries (extraction of coal, oil and natural gas, non-energy minerals). Indeed, the information on these industries provided by the Italian IO tables is not representative since only some parts of these industries are present in Italy. We therefore add three virtual industries to the matrices, drawing from Eurostat datasets from other European countries which represent the missing parts of these industries.

Domestic emissions (${}_{d}\mathbf{E}$) may be written as total emissions that would have occurred if all production steps had been carried out domestically (**E**) *minus* emission avoided due to (intermediate and final) imports (${}_{m}\mathbf{E}$):[5]

$$ {}_{d}\mathbf{E} = \mathbf{E} - {}_{m}\mathbf{E}. \tag{1} $$

In the following formulae we use **r** for the vector of GHG emission intensities of output by industry, ${}_{d}\mathbf{A}$ for the matrix of domestic production direct coefficients, ${}_{m}\mathbf{A}$ for the matrix of intermediate import direct coefficients, $\mathbf{A} = {}_{d}\mathbf{A} + {}_{m}\mathbf{A}$ for the direct coefficient matrix of total intermediate inputs, ${}_{d}\mathbf{Y}$ for the matrix of final uses of domestic products by delivering industry and by category of final demand, ${}_{m}\mathbf{Y}$ for the matrix of final uses of imported products by delivering industry and by category of final demand and $\mathbf{Y} = {}_{d}\mathbf{Y} + {}_{m}\mathbf{Y}$ for the matrix of total final uses by delivering industry and by category of final demand.

The global-oriented IO model which calculates the total emissions activated world-wide by the final demand for Italian and imported products is:

$$ \mathbf{E} = \left\langle \mathbf{r}(\mathbf{I} - \mathbf{A})^{-1} \right\rangle \mathbf{Y}; \tag{2} $$

the actual emissions of the Italian production system, seen from the perspective of activation by final demand for domestically produced goods and services, are reallocated to the vertically integrated industries according to the domestic flows of intermediate inputs:

$$ {}_{d}\mathbf{E} = \left\langle \mathbf{r}(\mathbf{I} - {}_{d}\mathbf{A})^{-1} \right\rangle {}_{d}\mathbf{Y}; \tag{3} $$

the emissions avoided due to final imports are calculated as:

$$ {}_{m}\mathbf{E}_{y} = \left\langle \mathbf{r}(\mathbf{I} - \mathbf{A})^{-1} \right\rangle {}_{m}\mathbf{Y}; \tag{4} $$

the emissions activated by final demand through intermediate imports are calculated as a residual:

$$ {}_{m}\mathbf{E}_{x} = \mathbf{E} - {}_{d}\mathbf{E} - {}_{m}\mathbf{E}_{y} = \left\langle \mathbf{r}\left[(\mathbf{I} - \mathbf{A})^{-1} - (\mathbf{I} - {}_{d}\mathbf{A})^{-1} \right] \right\rangle {}_{d}\mathbf{Y}. \tag{5} $$

Results

Results for GHG emissions of the Italian economic activities, based on elaborations on matrices at current prices, are shown in Table 6.1. Various indicators have been calculated which reflect different meanings of the Italian economy.

Between 1995 and 2007, the actual GHG emissions of Italian economic activities do not fluctuate much and at the end of the period are only 15.1 million tonnes (MT) higher (3.4 per cent). However, the total emissions calculated for total final demand (whose composition by kind of final demand is discussed below) grow much more (73.9 MT; 11.9 per cent); this is due to the dynamics of foreign trade: the emissions avoided due to imports increase by almost 58.8 MT (32.2 per cent). In particular, GHG emissions avoided due to intermediate imports activated by final demand are 41.3 MT (28.5 per cent) higher at the end of the period while those avoided due to final imports grow by 17.6 MT, a striking 45.9 per cent.

The emissions avoided due to imports as a whole amount to about 30 per cent of the total emissions from 1995 to 2003; their share increases up to 35 per cent in 2007, as if the increase of the emissions activated by exports in the same period (2003–2007, from 27 per cent to 31 per cent) were transferred abroad through imports. Indeed, the imports that enter as intermediate inputs in the Italian production processes form 75 per cent of total emissions avoided due to imports in 2003 and 77 per cent in 2007. The final (domestic and foreign) demand for Italian products activates on average in the period 93 per cent of total emissions; these emissions increase by 56.3 MT (9.6 per cent) from 1995 to 2007.

With regard to the composition by kind of final demand of the estimated total emissions, those activated by private and public consumption account on average in the period for 56 per cent, but decrease from 57 per cent in 2003 to 52 per cent in 2007; gross capital formation accounts for 16 per cent without remarkable changes whereas the remaining average 28 per cent, increasing up to 31 per cent in 2007, is ascribable to exports.

Table 6.1 also reports the balance of the emissions 'embodied' in trade, here defined as the GHG emissions virtually imported via the export of goods and services minus the GHG emissions virtually exported via the import of good and services (avoided emissions). A negative trend of the Italian balance is shown throughout the whole period; this demonstrates that the global emissions caused by the Italian production and consumption patterns are higher than actual emissions of Italian resident producers.

Although these results are affected by a price effect on emission estimates whose direction is unpredictable (inflation entails higher nominal values for final demand, but also lower direct emission intensities), they show that imports played an important role in keeping actual GHG emissions low.

Results by vertically integrated industry in 2007 and their 1995–2007 changes are shown in Table 6.2[6] (more detailed data by industry are available on request). Over the period, 'Manufacturing' industries (NACE divisions 15 to 37) are responsible for most of Italian actual emissions (42.4 per cent on average). In 2007, the actual emissions of this vertically integrated sector account for 185.1 MT and they decrease by 1.8 per cent from 1995. The total emissions of these industries increase up to 353.4 MT (and have grown by 12 per cent since 1995) because of the importance for 'Manufacturing' both of imports on the supply side (168.4 MT in 2007 and +32.6 per cent since 1995) and exports on the demand side (183.9 MT in 2007 and +26.8 per cent since 1995). Indeed, these economic activities account on average over the period for almost 70 per cent of emissions ascribable to final imports and activated through intermediate imports and for 83.5 per cent of total emissions ascribable to exports. The

Table 6.1 Italian GHG emissions ascribable to final demand, 1995–2007 (*million tonnes*)

	1995	*1996*	*1997*	*1998*	*1999*	*2000*	*2001*	*2002*	*2003*	*2004*	*2005*	*2006*	*2007*
Actual emissions of Italian production (a)	439.8	428.8	433.6	438.3	439.5	449.4	451.7	452.5	462.5	465.8	464.2	458.5	454.9
Emissions ascribable to final imports and activated through intermediate imports (m)	*183.0*	*161.0*	*173.3*	*175.9*	*173.3*	*201.3*	*204.9*	*197.6*	*200.1*	*211.8*	*217.5*	*233.0*	*241.8*
Of which													
Emissions ascribable to final demand for imported products (m^f)	**38.3**	**34.6**	**38.3**	**40.8**	**42.6**	**49.5**	**47.8**	**47.6**	**49.0**	**52.3**	**52.4**	**54.8**	**55.9**
Emissions activated through intermediate imports ($m^i = m - m^f$)	**144.6**	**126.4**	**135.0**	**135.1**	**131.1**	**151.8**	**157.1**	**150.0**	**151.0**	**159.4**	**165.0**	**178.2**	**185.9**
Total emissions ascribable to all final uses ($b = a + m$)	*622.8*	*589.7*	*606.9*	*614.2*	*613.2*	*650.6*	*656.6*	*650.1*	*662.6*	*677.5*	*681.6*	*691.5*	*696.7*

Of which													
Total emissions ascribable to Exports (x)	**173.7**	**162.2**	**169.7**	**167.0**	**160.8**	**186.9**	**189.5**	**178.2**	**178.9**	**192.2**	**199.5**	**211.1**	**219.4**
GHG footprint of domestic uses	**449.1**	**427.6**	**437.2**	**447.2**	**452.4**	**463.8**	**467.0**	**471.9**	**483.6**	**485.3**	**482.2**	**480.4**	**477.3**
Of which													
Total emissions ascribable to final consumption expenditure	***350.7***	***338.2***	***344.6***	***353.3***	***356.3***	***359.5***	***364.8***	***366.0***	***378.8***	***375.3***	***376.5***	***367.5***	***362.4***
Total emissions ascribable to gross capital formation	***98.3***	***89.3***	***92.6***	***94.0***	***96.1***	***104.3***	***102.3***	***105.9***	***104.8***	***110.0***	***105.7***	***112.9***	***114.9***
Emissions ascribable to the final demand for Italian products ($c=a+m^i$)	**584.4**	**555.1**	**568.6**	**573.4**	**570.6**	**601.2**	**608.8**	**602.5**	**613.5**	**625.2**	**629.2**	**636.7**	**640.8**
Balance of emissions embodied in trade ($x-m$)	**–9.3**	**1.2**	**–3.6**	**–8.9**	**–12.9**	**–14.4**	**–15.4**	**–19.4**	**–21.1**	**–19.5**	**–18.0**	**–21.9**	**–22.4**

analysis by kind of final demand of the results for the vertically integrated 'Manufacturing' activities shows that more than half of the 2007 emissions is triggered by foreign demand (183.9 MT and has increased by 26.8 per cent since 1995) and 48 per cent by domestic final uses (169.6 MT, not changing much since 1995). As far as the emissions ascribable to total imports and exports are concerned, crucial relevance among these industries must be assigned to the NACE sub-sections DA 'Manufacture of food products; beverages and tobacco', DB 'Manufacture of textiles and textile products', DG 'Manufacture of chemicals, chemical products and man-made fibres', DJ 'Manufacture of basic metals and fabricated metal products', DK 'Manufacture of machinery and equipment n.e.c.', DL 'Manufacture of electrical and optical equipment' and DM 'Manufacture of transport equipment'.

Attention should also be drawn to the service activities (NACE 50 to 95) and 'Construction' – respectively 30.6 per cent and 7.4 per cent of Italian total emissions on average during the period – which are characterized by a higher share of emissions activated through intermediate imports than the share of emissions ascribable to final demand for imported products. These vertically integrated activities show different characteristics on the demand side: services are more connected to final private and public consumption and, though to a lesser extent, to exports; 'Construction' emissions are basically and obviously related to gross capital formation.

The agricultural sector (NACE 01–05) data shows that the emissions avoided due to imports are mainly ascribable to final demand for imported products (5 MT in 2007, +20.7 per cent from 1995). Emissions avoided due to this sector's intermediate imports are quite low (0.9 MT); it can therefore be concluded that the emissions virtually exported by selling Italian agricultural products abroad are indeed actual emissions of Italian producers and not the result of a transfer of virtually exported emissions.

The structural decomposition analysis of total GHG emissions 'embodied' in Italian industries' final products and imports

The model

The decomposition approach builds on the general extension of the Dietzenbacher and Los (1998) method suggested by Seibel (2003) which is equally suited for use both with the IO model and with sector-level data.

If we also consider the composition of total final demand by industry, equation (2) can be written as:

$$\mathbf{e}' = \left\langle \mathbf{r}(\mathbf{I}-\mathbf{A})^{-1} \right\rangle \left\langle \mathbf{c} \right\rangle \mathbf{y}' \tag{6}$$

where the vector $\mathbf{c}' = \mathbf{y}'y^{-1}$ describes the composition of total final demand by delivering industry and $\mathbf{y}'$ is the vector of total final uses by delivering industry.

In order to break down the change in emissions in the period $(t)-(t+1)$, the basic idea of the decomposition analysis is to split $\mathbf{e}'$ in the changes, *ceteris paribus*, of the four components identified in equation (6): 'emission intensity' and the 'Leontief' components, which basically are two technological factors; the 'composition of final demand' by delivering industry (one euro spent on final product i activates a different quantity of emissions than one euro spent on final product j); the level of total final demand which is the most significant driving force in emission growth. We calculated the changes in the total emissions ascribable to each of the above-mentioned determinants on a year-by-year basis for the period 1999–2007 (due to the availability of supply-and-use tables evaluated at constant prices of the previous year). We did this by subtracting the emissions calculated for the first year using monetary data at current prices from the emissions calculated for the second year by using data at the previous year's prices. These changes were then cumulated in order to reconstruct the evolution of the total change and its components in the whole period.

Results

Figure 6.1 shows the results of this exercise. The change in total emissions in the period 1999–2007, thus calculated, was of about 46 MT, or 8 per cent of the initial value. This increase is exclusively due to the 'total final demand' component, which alone would have implied 101 MT more emissions, reflecting the increasing level of economic activities during the period. The other components were either on the whole insignificant ('input structure' or 'Leontief' component, and 'composition of final demand') or counterbalanced this effect ('direct emission intensity' which would have entailed a 55.5 MT reduction in GHG emissions).

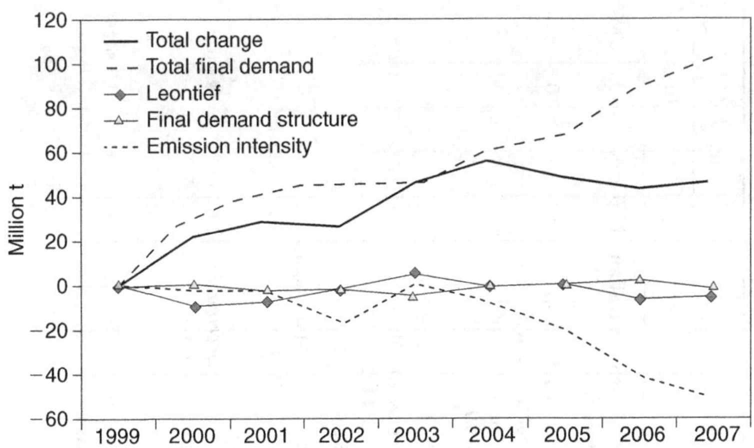

Figure 6.1 Cumulative changes between 1999 and 2007 in total emissions ascribable to Italian production and final uses, broken down by effect (million tonnes).

Table 6.2 Italian GHG emissions by vertically integrated industry in 2007 and 1995–2007 changes *(million tonnes and percentages)*

	Industries (NACE Rev. 1.1)							
	01–05	*10–14*	*15–37*	*40–41*	*45*	*50–55*	*60–63*	*64–95*
Actual emissions of Italian production (a)	15.0 6.9%	0.6 -0.6%	185.1 -1.8%	43.9 -10.1%	41.3 17.8%	75.7 12.6%	33.2 14.7%	60.1 6.2%
Emissions ascribable to final imports and activated through intermediate imports (m)	5.9 18.5%	0.1 5.0%	168.4 32.6%	4.9 -4.6%	12.3 21.3%	23.8 33.7%	8.3 41.7%	18.1 52.0%
Of which								
Emissions ascribable to final demand for imported products (m^f)	5.0 20.7%	0.02 31.2%	46.7 47.3%	0.04 -59.5%	0.00 -21.3%	1.3 43.2%	2.3 113.4%	0.5 41.4%
Emissions activated through intermediate imports ($m^i = m - m^f$)	0.9 7.6%	0.1 1.7%	121.7 27.6%	4.9 -3.5%	12.3 21.3%	22.5 33.2%	5.9 25.2%	17.6 52.3%
Total emissions ascribable to all final uses (b=a+m)	20.9 10.0%	0.7 0.2%	353.4 12.0%	48.8 -9.6%	53.7 18.5%	99.5 17.1%	41.5 19.2%	78.2 14.1%
Of which								
Total emissions ascribable to Exports (x)	6.5 18.9%	0.7 50.1%	183.9 26.8%	1.0 1.6%	0.4 10.4%	10.5 26.9%	11.9 8.8%	4.4 113.1%

GHG footprint of domestic uses	14.4 6.4%	0.02 –90.3%	169.6 –0.5%	47.8 –9.8%	53.2 18.6%	88.9 16.0%	29.6 24.0%	73.8 11.0%
Of which								
Total emissions ascribable to final consumption expenditure	13.8 6.1%	0.1 38.2%	119.8 –5.2%	46.8 –7.4%	3.0 –110.8%	81.2 14.0%	27.4 22.6%	70.3 10.2%
Total emissions ascribable to gross capital formation	0.6 12.6%	–0.1 –150.2%	49.7 13.0%	0.9 –61.2%	50.2 21.0%	7.7 42.5%	2.2 45.3%	3.5 29.0%
Emissions ascribable to the final demand for Italian products ($c=a+m^i$)	15.9 7.0%	0.7 –0.3%	306.7 8.1%	48.7 –9.5%	53.7 18.5%	98.2 16.8%	39.2 16.2%	77.7 14.0%
Balance of emissions embodied in trade ($x-m$)*	0.5 0.4	0.6 0.3	15.5 18.0	–3.9 –4.2	–11.9 –9.8	–13.2 –9.5	3.7 5.1	–13.7 –9.8

Notes
* In this line 1995 values are shown instead of percentage changes.
Legend
Description of the NACE codes: (01–05) ‘Agriculture, forestry and fishing’; (10–14) ‘Mining and extraction’; (15–37) ‘Manufacturing’; (40–41) ‘Electricity, gas and water’; (45) ‘Construction’; (50–55) ‘Trade’; (60–63) ‘Transport’; (64–95) ‘Services’.

The structural decomposition analysis of total GHG emissions avoided due to final and intermediate imports

With regard to emissions triggered by imported commodities (${}_{\mathbf{m}}\mathbf{e}_{\mathbf{y}}$), we test the hypothesis that significant avoidance of emissions is due to incremental displacement of production by means of two separate SDA, one for final and one for intermediate imports.

Final imports: the model

In order to evaluate this 'displacement' hypothesis for final imports for each year and industry, we take the previous year's imports share on total final demand for the products of that industry as reference so that if this share does not change, nothing is ascribed to the displacement component, *even though the imports may have changed in absolute terms*. We therefore use the following model in which the SDA follows the methodology introduced in the previous section:

$$ {}_{\mathbf{m}}\mathbf{e}_{\mathbf{y}} = \left\langle \mathbf{r}(\mathbf{I}-\mathbf{A})^{-1} \right\rangle \left\langle \mathbf{s} \right\rangle \mathbf{y}' \tag{7} $$

where $\langle\mathbf{s}\rangle = \langle{}_{\mathbf{m}}\mathbf{y}\rangle\langle\mathbf{y}\rangle^{-1}$ denotes the displacement component: this is the vector of the imported final demand's share of total final demand by the delivering industry; $\mathbf{y}'$ represents the vector of total final uses by the delivering industry. The 'emission intensity' and the 'Leontief' components were already identified in equation (6).

Intermediate imports: the model

With regard to GHG emissions activated by final demand through intermediate imports, the formula used for the SDA is the one already specified in equation (5). This expression comprises the matrix ${}_{\mathbf{d}}\mathbf{A}$ whose elements are[7] ${}_{d}a_{ij} = a_{ij} - {}_{m}a_{ij}$. To identify the 'displacement' component, we now aim to single out, in the change ${}_{d}a_{ij}$, how much is due to the technical factor a_{ij} and how much is due to the shift abroad of the source of the intermediate inputs supplied by industry i to industry j.

Let g be the percentage change of a_{ij} in the period $(t)-(t+1)$ and let us calculate the value ${}_{d}\hat{a}_{ij}^{(t+1)}$ that ${}_{d}a_{ij}$ would have at time $(t+1)$ if it changed in the same way as a_{ij}:

$$ {}_{d}\hat{a}_{ij}^{(t+1)} = {}_{d}a_{ij}^{(t)} + g\,{}_{d}a_{ij}^{(t)} = {}_{d}a_{ij}^{(t)} a_{ij}^{(t+1)} / a_{ij}^{(t)} \tag{8} $$

The difference between this coefficient and the actual value ${}_{d}a_{ij}^{(t+1)}$ can be conceptualized as the effect of displacement of supply sources in favour of foreign activities. On this basis, we can now study the year-by-year changes in ${}_{\mathbf{m}}\mathbf{e}_{\mathbf{x}}$ for testing the effect on the avoided emissions of the displacement of intermediate production. The change in the period $(t)-(t+1)$ will be:

$$
\begin{aligned}
\Delta\,{}_{m}\mathbf{e}_{x} &\equiv {}_{m}\mathbf{e}_{x}^{(t+1)} - {}_{m}\mathbf{e}_{x}^{(t)} \\
&\equiv \left[{}_{m}\mathbf{e}_{x}^{(t+1)} - {}_{m}\underset{\cdot}{\mathbf{e}}_{x}^{(t+1)}\right] + \left[{}_{m}\underset{\cdot}{\mathbf{e}}_{x}^{(t+1)} - {}_{m}\underset{\cdot}{\hat{\mathbf{e}}}_{x}^{(t+1)}\right] \\
&+ \left[{}_{m}\underset{\cdot}{\hat{\mathbf{e}}}_{x}^{(t+1)} - {}_{m}\underset{\cdot}{\hat{\mathbf{e}}}{}^{*}{}_{x}^{(t+1)}\right] + \left[{}_{m}\underset{\cdot}{\hat{\mathbf{e}}}{}^{*}{}_{x}^{(t+1)} - {}_{m}\mathbf{e}_{x}^{(t)}\right]
\end{aligned}
\tag{9}
$$

where we define ${}_{m}\mathbf{e}_{x}$ as the vector of emissions activated by final demand through intermediate imports, calculated using tables at current prices, by delivering industry, ${}_{m}\underset{\cdot}{\mathbf{e}}_{x}$ as the vector of emissions activated through intermediate imports, valued at previous year prices, by delivering industry, ${}_{m}\underset{\cdot}{\mathbf{e}}_{x}$ as the vector of emissions that would have been produced if the domestic share of intermediate inputs had remained unchanged, i.e. ${}_{m}\underset{\cdot}{\mathbf{e}}_{x}$ assuming g as the percentage change of ${}_{d}a_{ij}$ and ${}_{m}\underset{\cdot}{\hat{\mathbf{e}}}{}^{*}{}_{x}$ as the vector of emissions ${}_{m}\underset{\cdot}{\hat{\mathbf{e}}}_{x}$ with ${}_{d}\mathbf{y}$ valued at current prices.

The change in the emissions activated through intermediate imports is broken down into four components. The first element in equation (9) shows the change due to the change in prices, the second, the displacement effect, and the last two components explain the change due to other factors which are not linked to delocalization, namely domestic final demand and technology.

Here we are interested in the second component of the above identity which shows the change in total emissions that would have occurred if the domestic intermediate input matrix had remained unchanged (and therefore also the intermediate import matrix).

Results: the role of displacement from decomposition analyses

Table 6.3 summarizes the results of the two decomposition analyses. It can be noted that actual domestic GHG emissions from production would have grown much more than they did (38.1 MT rather than 15.3 MT) if there had been no displacement of production. Indeed, the change in composition of the sources of supply was such that only around one-fourth of the increase of the emissions avoided due to imports (30.8 MT) was due to changes affecting the total size of the emissions of the vertically integrated industries (8 MT due to the total final demand, intermediate inputs needs and emission intensity of output) whereas the remaining three-fourths may be ascribed to a displacement effect.

Table 6.3 Effect of displacement of production on GHG emissions: Italy, 1999–2007 (million tonnes)

	Actual domestic emissions from production	*Emissions avoided thanks to imports*			*Total emissions*
		Intermediate imports	*Final imports*	*Total*	
Displacement	–22.8	16.6	6.2	22.8	–
Changes affecting total emissions	38.1	5.4	2.6	8.0	46.1
1999–2007 change	15.3	22.0	8.8	30.8	46.1

Figure 6.2 shows how the two components of imports contributed to the overall change in the period by reporting the cumulated effect of production displacement. It may be noted that there was no increase in the total, i.e. no displacement effect, from 2000 to 2001, whereas a definite trend emerges for the whole subsequent period in which two different sub-periods may be identified, one (2001–2004) where displacement grows due to final imports and another (2004–2007) where it grows due to intermediate imports.

Figure 6.3 shows the role of displacement at the level of the vertically integrated industries from 1999 to 2007 for the 14 'Manufacturing' NACE subsections and four other activities whose results are worth taking into account.

We find most of these vertically integrated industries in the first quadrant of the diagram which indicates a shift abroad both of final demand and intermediate consumption. In particular, it is important to note that the latter are activated by the part of final demand which was not displaced. Within this quadrant, 'Manufacture of chemicals, chemical products and man-made fibres' (DG), 'Manufacture of textiles and textile products' (DB), 'Wholesale and retail trade' and 'Hotels and restaurants' (G and H) and, though to a lesser extent, 'Manufacture of rubber and plastic products' (DH) and 'Manufacture of pulp, paper and paper products; publishing and printing' (DE) show a preponderance in displacing the intermediate steps of production processes; on the contrary, the displacement of final imports prevails in 'Manufacture of transport equipment' (DM).

A sort of replacement occurs both for agricultural and land transport activities. With regard to the fourth quadrant, 'Manufacture of machinery and equipment n.e.c.' (DK) and 'Manufacture of basic metals and fabricated metal products' (DJ) only show a shift in intermediate inputs.

Conclusions

Although not huge in relative terms ('only' 5.2 per cent of the 1999 level of domestic GHG emissions from production), there was a significant displacement

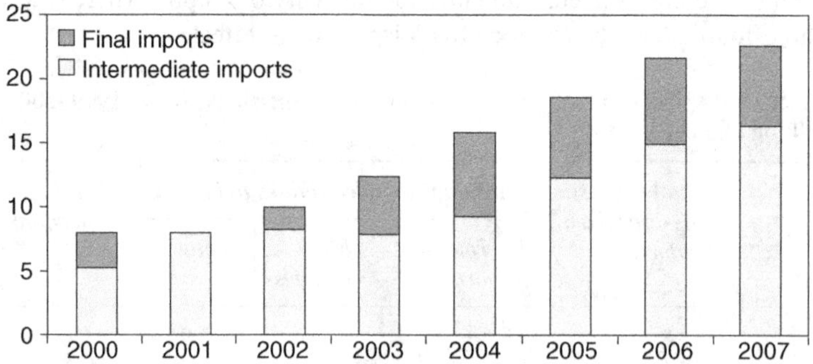

Figure 6.2 Cumulated effect of production displacement through intermediate and final imports, 1999–2007 (million tonnes).

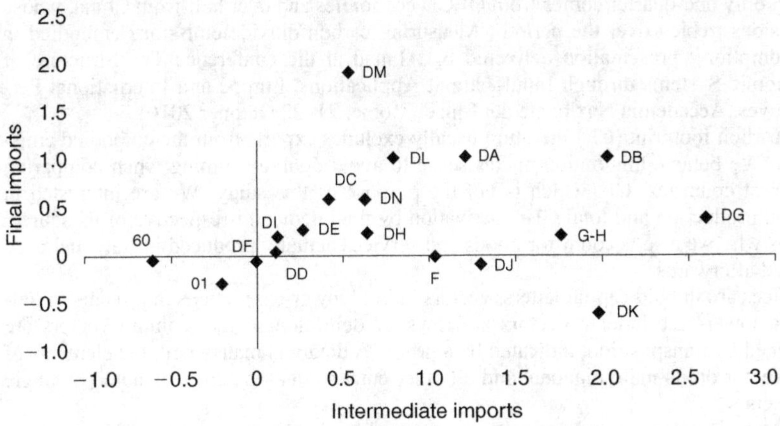

Figure 6.3 Displacement effect on emissions by type of import and by vertically integrated industry, 1999–2007 (million tonnes).

Notes
Description of the NACE codes: (01) 'Agriculture, hunting and related service activities'; (DA) 'Manufacture of food products; beverages and tobacco'; (DB) 'Manufacture of textiles and textile products'; (DC) 'Manufacture of leather and leather products'; (DD) 'Manufacture of wood and wood products'; (DE) 'Manufacture of pulp, paper and paper products; publishing and printing'; (DF) 'Manufacture of coke, refined petroleum products and nuclear fuel'; (DG) 'Manufacture of chemicals, chemical products and man-made fibres'; (DH) 'Manufacture of rubber and plastic products'; (DI) 'Manufacture of other non-metallic mineral products'; (DJ) 'Manufacture of basic metals and fabricated metal products'; (DK) 'Manufacture of machinery and equipment n.e.c.'; (DL) 'Manufacture of electrical and optical equipment; (DM) 'Manufacture of transport equipment'; (DN) 'Manufacturing n.e.c.'; (F) 'Construction'; (G and H) 'Wholesale and retail trade' and 'Hotels and restaurants'; (60) 'Land transport'.

of the Italian GHG emissions towards the rest of the world, stemming mainly from the growing shares of imports in the final demand for products of the manufacturing industries and in the intermediate demand of these industries. This means that emissions avoided due to imports grew more rapidly than total final demand due to a shift in composition of supply of final and intermediate products towards the rest of the world. This indicates the importance of also analysing the total environmental flows of economies and going beyond territorial estimations of environmental pressures. If they really care about the global climate and, more generally, the global environment, policy-makers should be aware that, although hidden under some foreign carpet, this dirt may be known, measured and considered a target in global and national policies.

Notes

1 Tudini and Vetrella, in Chapter 1, provide a comprehensive picture of NAMEA-type accounts.

2 It also makes sense to pose the question at domestic level to address 'Sustainable Consumption and Production' policies, for example.

3 Ahmad writes that of the 40 per cent increase in CO_2 emissions between 1990 and 2007, only one-quarter comes from OECD economies and over half from China, whose emissions trebled over the period ('Measuring carbon dioxide emissions embodied in consumption': presentation delivered by Ahmad at the conference The Structure of Economic Systems through Input–Output Applications: Europe and International Perspectives, Accademia Nazionale dei Lincei, Rome, 21–22 October 2010).

4 The carbon footprint (CF) literature usually excludes exports from the embodied emissions. We believe this option makes sense to avoid double counting when comparing different countries' CF, which is not the purpose of this study. We are interested in Italian production and total GHG activation by final demand irrespective of its source; this is why we also account for goods and services actually produced in Italy and used abroad afterwards.

5 Matrices are in bold capital letters, vectors in bold lower-case letters and scalars in italicized lower-case letters. Vectors are rows by definition so that column vectors are obtained by transposition, indicated by a prime. A diagonal matrix with the elements of any vector on its main diagonal and all other entries equal to zero is denoted by angle brackets < >.

6 For analysis purposes, industries are aggregated into eight macro sectors. Note that, in order to show more homogeneous activities as for environmental pressures, the NACE section 'Transport' (60–63) does not constitute the 'Post and telecommunications' division which is included in the aggregation 'Services' (64–95).

7 The index i is the delivering industry and j the receiving industry.

References

Costa, S. and Ferri, G. (2007) 'The determinants and the employment effects of international outsourcing: the case of Italy', Department of Economics and Mathematics, University of Bari, Working Paper no. 16.

Dietzenbacher, E. and Los, B. (1998) 'Structural decomposition techniques: sense and sensitivity', *Economic Systems Research*, 10: 307–323.

Dinda, S. (2004) 'Environmental Kuznets curve hypothesis: a survey', *Ecological Economics*, 49: 431–455.

Eurostat (2008) 'Eurostat manual of supply, use and input–output tables', Luxembourg: Eurostat.

Femia, A. and Marra Campanale, R. (2010) 'Production-related air emissions: a decomposition analysis for Italy', in M. Mazzanti and A. Montini (ed.), *Environmental Efficiency, Innovation and Economic Performances*, London: Routledge.

Femia, A. and Panfili, P. (2005) 'Analytical applications of the NAMEA', paper presented at the annual meeting of the Italian Statistics Society, Rome: Italian Statistics Society.

ISTAT (2010) *Air emission accounts*. Online: http://www.istat.it/en/ (accessed 15 December 2010).

ISTAT (2011) *Il sistema delle tavole input–output*. Online: http://www.istat.it/en/ (accessed 4 January 2011).

Minx, J., Wiedmann, T., Wood, R., Peters, G., Lenzen, M., Owen, A., Scott, K., Barrett, J., Hubacek, K., Baiocchi, G., Paul, A., Dawkins, E., Briggs, J., Guan, D., Suh, S. and Ackermann, F. (2009) 'Input–output analysis and carbon footprinting: an overview of regional and corporate applications', *Economic Systems Research*, 21: 187–216.

OECD (2007) *Offshoring and Employment: Trends and Impacts*, Paris: OECD.

Pasinetti, L. (1973) 'The notion of vertical integration in economic analysis', *Metroeconomica*, 25: 1–29.

Peters, G. and Hertwich, E. (2008) 'CO_2 embodied in international trade with implications for global climate change policy', *Environmental Science and Technology*, 42: 1401–1407.

Rørmose, P., Olsen, T. and Hansen, D. (2010) *GHG Emissions Embodied in Trade*, Statistics Denmark Technical Report to Eurostat, Luxembourg: Eurostat.

Seibel, S. (2003) 'Decomposition analysis of carbon dioxide emission changes in Germany: conceptual framework and empirical results', Eurostat Working Papers and Studies, Luxembourg: Eurostat.

Weinzettel, J. and Kovanda, J. (2009) 'Assessing socioeconomic metabolism through hybrid life cycle assessment: the case of the Czech Republic', *Journal of Industrial Ecology*, 13: 607–621.

Part II

NAMEA and input–output frameworks

Integration, analyses and policy issues in a European perspective

7 Comparisons of the European carbon footprint (2000–2006) from three different perspectives within a multi-regional framework

José Manuel Rueda-Cantuche

Introduction

We would expect the levels of carbon dioxide emissions per unit of production in non-European Union (EU) countries to be much greater than those in EU countries. Accordingly, a reduction in the import shares of intermediate inputs and/or final demand of the European economy should lead to more domestic intermediate and final uses and, therefore, to a reduction in the carbon footprint of the European economy when compared with the initial situation. However, we will prove that only when different emission coefficients are considered for non-EU and EU countries, then the European import shares reduction that occurred during the period 2000–2006 will be fully captured by carbon footprint analysis. Otherwise, the derived carbon footprint will be distorted by the fact that the evolution of the import shares will not affect those estimations. Moreover, the results of our analysis are independent of the assumption of common production technologies in the rest of the world.

Carbon footprint analysis within a multi-regional framework is not new in the literature (Serrano and Dietzenbacher, 2010) but nevertheless, the lack of available data has not led to a comprehensive formalization of the calculation of the carbon footprint but only to the establishment of ad-hoc models that reflect their actual data requirements very well. In this chapter, we will formalize the calculation of the carbon footprint within a multi-regional input–output framework of two regions, thus providing the corresponding expressions to be used just in case full information is available on all regions. For the sake of clarification, we will refer hereafter to carbon footprint in the sense that it will only include indirect emissions and not those directly derived from the use of goods and services by households (e.g. driving, cooking, heating).

From the general expression derived for the calculation of the carbon footprint, we will derive two additional similar simplified expressions by making a set of nested assumptions, all of which will assume domestic and/or total (domestic plus imports) technologies for non-EU countries. The empirical analysis of the three approaches will show us in particular that the assumption of domestic emission coefficients in the rest of the world can not only seriously distort the derived evolution of the carbon footprint in Europe during the period

2000–2006 but also reach on average up to 7.2 per cent lower estimated emission levels during the same period. The underlying reasons for this behaviour are two-fold: (a) non-EU emission coefficients are generally much greater than those in the EU; (b) the shifts in the import shares of intermediate and/or final uses are not at all captured by a model in which industry-based emissions per unit of production are equalized for all countries.

The next section provides a generalized multi-regional framework for the calculation of the carbon footprint in the two regions specified in the model (i.e. in our case, EU and non-EU countries) under the assumption that we have all the necessary information. Subsequent assumptions will be incorporated, leading to simplified expressions. Then, I describe the data sources and analyse the results for the EU27 during the period 2000–2006. I conclude with the main lessons learnt from this chapter for future carbon footprint analysis.

Methodology

Let us consider a world with only two regions (u=European Union; r=rest of the world) that may differ in production technologies and/or emission levels per unit of production. Each one of the regions has economic activities classified into n industries, all of which can produce more than one product that might be used (abroad or domestically) by other industries as intermediate input and/or capital formation and by households, non-profit organizations and the government with final consumption purposes. The Leontief quantity model in its partitioned form is given by[1] (Serrano and Dietzenbacher, 2010; Dietzenbacher, 2002):

$$\begin{bmatrix} x_u \\ x_r \end{bmatrix} = \begin{bmatrix} A^{uu} & A^{ur} \\ A^{ru} & A^{rr} \end{bmatrix} \begin{bmatrix} x_u \\ x_r \end{bmatrix} + \begin{bmatrix} y_{uu} + y_{ur} \\ y_{ru} + y_{rr} \end{bmatrix} \quad (1)$$

with A^{uu} and A^{rr} being the matrices of domestic technical coefficients of EU and non-EU countries, respectively and A^{ru} and A^{ur} as the matrices of imported input coefficients for the EU and the non-EU countries, respectively. The same applies to the different components of final demand (y). The total industry outputs are given by the vectors x_u and x_r. In its partitioned form, the solution to the Leontief system is:

$$\begin{bmatrix} x_u \\ x_r \end{bmatrix} = \left(\begin{bmatrix} I & 0 \\ 0 & I \end{bmatrix} - \begin{bmatrix} A^{uu} & A^{ur} \\ A^{ru} & A^{rr} \end{bmatrix} \right)^{-1} \begin{bmatrix} y_{uu} + y_{ur} \\ y_{ru} + y_{rr} \end{bmatrix}$$

$$\begin{bmatrix} x_u \\ x_r \end{bmatrix} = \begin{pmatrix} I - A^{uu} & -A^{ur} \\ -A^{ru} & I - A^{rr} \end{pmatrix}^{-1} \begin{bmatrix} y_{uu} + y_{ur} \\ y_{ru} + y_{rr} \end{bmatrix}$$

$$\begin{bmatrix} x_u \\ x_r \end{bmatrix} = \begin{pmatrix} L^{uu} & L^{ur} \\ L^{ru} & L^{rr} \end{pmatrix} \begin{bmatrix} y_{uu} + y_{ur} \\ y_{ru} + y_{rr} \end{bmatrix} \quad (2)$$

A simplified version of the expressions for the calculation of the elements of the partitioned Leontief inverse (L) can be found in Dietzenbacher (2002, p. 129).

For the calculation of the emissions associated with EU and non-EU countries, let us assume that each region emits c_u and c_r units of a pollutant (e.g. carbon dioxide) per unit of production. These vectors are of order n (number of industries). Therefore, the vector of total emissions produced in each region, denoted as v_u and v_r, will be given by:

$$\begin{bmatrix} v_u \\ v_r \end{bmatrix} = \begin{bmatrix} \hat{C}_u & 0 \\ 0 & \hat{C}_r \end{bmatrix} \begin{bmatrix} x_u \\ x_r \end{bmatrix} = \begin{bmatrix} \hat{C}_u & 0 \\ 0 & \hat{C}_r \end{bmatrix} \begin{pmatrix} L^{uu} & L^{ur} \\ L^{ru} & L^{rr} \end{pmatrix} \begin{bmatrix} y_{uu} + y_{ur} \\ y_{ru} + y_{rr} \end{bmatrix} \quad (3)$$

which properly extended is:

$$\begin{bmatrix} \hat{C}_u & 0 \\ 0 & \hat{C}_r \end{bmatrix} \begin{pmatrix} L^{uu} & L^{ur} \\ L^{ru} & L^{rr} \end{pmatrix} \begin{bmatrix} y_{uu} \\ y_{ru} \end{bmatrix} + \begin{bmatrix} \hat{C}_u & 0 \\ 0 & \hat{C}_r \end{bmatrix} \begin{pmatrix} L^{uu} & L^{ur} \\ L^{ru} & L^{rr} \end{pmatrix} \begin{bmatrix} y_{ur} \\ y_{rr} \end{bmatrix}$$

$$= \begin{pmatrix} \hat{C}_u L^{uu} & \hat{C}_u L^{ur} \\ \hat{C}_r L^{ru} & \hat{C}_r L^{rr} \end{pmatrix} \begin{bmatrix} y_{uu} \\ y_{ru} \end{bmatrix} + \begin{pmatrix} \hat{C}_u L^{uu} & \hat{C}_u L^{ur} \\ \hat{C}_r L^{ru} & \hat{C}_r L^{rr} \end{pmatrix} \begin{bmatrix} y_{ur} \\ y_{rr} \end{bmatrix}$$

$$= \begin{pmatrix} \hat{C}_u L^{uu} y_{uu} + \hat{C}_u L^{ur} y_{ru} \\ \hat{C}_r L^{ru} y_{uu} + \hat{C}_r L^{rr} y_{ru} \end{pmatrix} + \begin{pmatrix} \hat{C}_u L^{uu} y_{ur} + \hat{C}_u L^{ur} y_{rr} \\ \hat{C}_r L^{ru} y_{ur} + \hat{C}_r L^{rr} y_{ru} r \end{pmatrix}$$

$$= \begin{pmatrix} v_{uu} \\ v_{ru} \end{pmatrix} + \begin{pmatrix} v_{ur} \\ v_{rr} \end{pmatrix} = \begin{bmatrix} v_u \\ v_r \end{bmatrix} \quad (4)$$

with the following definitions given below in points (a) to (h): (a) v_{uu}: emissions produced in the EU derived from the EU final demand of domestically produced commodities (e.g. purchase of a German computer by a European resident); plus the emissions produced in the EU for the production of an exported commodity that will be used by the rest of the world to produce something else that the EU will import (e.g. exports of EU electronic components for the production of Japanese computers that will be imported by the EU). The former can be considered an intra-regional effect ($\hat{C}_u L^{uu} y_{uu}$) whereas the latter can be considered an inter-regional effect ($\hat{C}_u L^{ur} y_{ru}$), both occurring within the EU; (b) v_{ru}: emissions produced in non-EU countries derived from the imported inputs needed to satisfy the EU final demand of domestically produced commodities (e.g. purchase of a German computer by an European resident that involves imports of electronic components from China); plus the emissions produced in the non-EU countries to satisfy the EU final demand of foreign commodities (e.g. imports of Japanese computers by the EU residents). Similarly, the former can be considered an inter-regional ($\hat{C}_r L^{ru} y_{uu}$) effect whereas the latter can be considered an intra-regional effect ($\hat{C}_r L^{rr} y_{ru}$), both occurring outside the EU; (c) $cf_u = v_{uu} + v_{ru}$: the sum of the two latter expressions yield the so-called carbon footprint (for carbon dioxide) which is the total emissions generated as a result of the intermediate and final demand of EU residents, independently of where the commodities were produced; (d) v_{rr}: emissions produced in non-EU countries derived from the non-EU final demand of their domestically produced commodities (e.g. purchase of a Japanese computer by a Chinese or another Japanese resident), plus the emissions produced in the rest of the world for the production of an exported commodity that will be used by the EU to produce something else that the rest of the world will import (e.g. exports of Chinese electronic components for the production of German computers that will be imported by the rest of the world). The former can be considered an intra-regional effect ($\hat{C}_r L^{rr} y_{rr}$) whereas the latter can be considered an inter-regional effect ($\hat{C}_r L^{ru} y_{ur}$), both occurring within the non-EU countries; (e) v_{ur}: emissions produced in EU countries derived from the imported inputs of the rest of the world needed to satisfy their own final demand of domestically produced commodities (e.g. purchase of a Japanese computer by a Chinese or another Japanese resident that involves imports of electronic components from the EU), plus the emissions produced in the EU countries to satisfy the non-EU final demand of European commodities (e.g. imports of German computers by the non-EU residents) – similarly, the former can be considered an inter-regional effect ($\hat{C}_u L^{ur} y_{rr}$) whereas the latter can be considered an intra-regional effect ($\hat{C}_u L^{uu} y_{ur}$), both occurring within the EU or outside the non-EU countries; (f) $cf_r = v_{ur} + v_{rr}$: the sum of the last two expressions now yield the estimated carbon footprint (for carbon dioxide) of the rest of the world which is the total emissions generated as a result of the intermediate and final uses of non-EU residents, independently of where the commodities were produced; (g) $v_u = v_{uu} + v_{ur}$: total emissions produced in the EU to satisfy domestic and foreign final demand and/or intermediate uses; (h) $v_r = v_{ru} + v_{rr}$: total emissions produced in non-EU countries (rest of the world) to satisfy their domestic and foreign final demand and/or intermediate uses.

For the interested reader, the so-called trade emission balance (TEB) or responsibility emission balance (REB) (Serrano and Dietzenbacher, 2010) can be calculated here by the difference between the emissions actually produced in the EU (v_u) and the European carbon footprint (cf_u) which turns out to be: $v_{ur} - v_{ru}$. However, a more detailed analysis of the two kinds of emission balance goes beyond the scope of this chapter. Hereafter, we will just focus our attention on the calculation of carbon footprints, and specifically that of the EU, and the discussion of proper measures for their analysis during the period 2000–2006.

From (4), the calculation of the carbon footprint for the European Union will be given by:

$$cf_u = \begin{pmatrix} 1 & 1 \end{pmatrix} \begin{pmatrix} v_{uu} \\ v_{ru} \end{pmatrix} = \hat{C}_u L^{uu} y_{uu} + \hat{C}_u L^{ur} y_{ru} + \hat{C}_r L^{ru} y_{uu} + \hat{C}_r L^{rr} y_{ru} \tag{5}$$

and for the rest of the world,

$$cf_r = \begin{pmatrix} 1 & 1 \end{pmatrix} \begin{pmatrix} v_{ur} \\ v_{rr} \end{pmatrix} = \hat{C}_u L^{uu} y_{ur} + \hat{C}_u L^{ur} y_{rr} + \hat{C}_r L^{ru} y_{ur} + \hat{C}_r L^{rr} y_{rr} \tag{6}$$

Unfortunately, the non-EU countries' final demand for their domestically produced commodities (y_{rr}) is not generally available and we will therefore only concentrate on the European carbon footprint. Two new ongoing EU-funded projects, EXIOPOL (Tukker *et al.*, 2009) and WIOD (Erumban *et al.*, 2010), are constructing international supply–use and world input–output tables which will definitely bring refreshing new data with which to use the above general expressions.

As will be shown later on in the empirical analysis, one key issue of carbon footprint analysis will be the consideration of non-EU emission coefficients as equal to that of the domestic values for the EU. Accordingly, the EU carbon footprint (2) turns into:

$$\begin{aligned} cf_u &= \begin{pmatrix} 1 & 1 \end{pmatrix} \begin{pmatrix} v_{uu} \\ v_{ru} \end{pmatrix} = \hat{C}_u L^{uu} y_{uu} + \hat{C}_u L^{ur} y_{ru} + \hat{C}_u L^{ru} y_{uu} + \hat{C}_u L^{rr} y_{ru} \\ &= \hat{C}_u \left(L^{uu} + L^{ru} \right) y_{uu} + \hat{C}_u \left(L^{ur} + L^{rr} \right) y_{ru} = \hat{C}_u \left[\left(L^{uu} + L^{ru} \right) y_{uu} + \left(L^{ur} + L^{rr} \right) y_{ru} \right] \end{aligned} \tag{7}$$

and the non-EU countries carbon footprint becomes:

$$\begin{aligned} cf_r &= \begin{pmatrix} 1 & 1 \end{pmatrix} \begin{pmatrix} v_{ur} \\ v_{rr} \end{pmatrix} = \hat{C}_u L^{uu} y_{ur} + \hat{C}_u L^{ur} y_{rr} + \hat{C}_u L^{ru} y_{ur} + \hat{C}_u L^{rr} y_{rr} \\ &= \hat{C}_u \left(L^{uu} + L^{ru} \right) y_{ur} + \hat{C}_u \left(L^{ur} + L^{rr} \right) y_{rr} = \hat{C}_u \left[\left(L^{uu} + L^{ru} \right) y_{ur} + \left(L^{ur} + L^{rr} \right) y_{rr} \right] \end{aligned} \tag{8}$$

However, there is another standard approach to calculating the carbon footprint of one region (e.g. the EU) which is very frequently used when there is incomplete information on production technology in the rest of the world and its final demand of either imported or domestically produced commodities. This approach consists of pre-multiplying the Leontief inverse matrix of total (import and domestic) input coefficients by the domestic emission coefficients and subsequently, post-multiplying the latter result by the total final demand of commodities by EU residents (either of domestically produced commodities or imported goods and services). This approach appears not to use any multi-regional model but we will prove that this is not the case. Basically, this method of calculating the carbon footprint of a region makes certain assumptions that simply lead to a more simplified expression in terms of the available data. These assumptions are as follows (mathematical proofs can be found in the Appendix): (a) *domestic emission coefficients applied abroad*; the amount of emissions per unit of commodity output is common to EU and non-EU countries. Obviously, it is implicit that the emission coefficients of the non-EU countries are equalized to those of the EU countries; (b) *changes in the shares of final demand and intermediate uses over the total commodity output of each region*; the two sides of the coin are that the rest of the world will not import commodities for intermediate uses from the EU but only for final demand uses and, subsequently, the non-EU countries will shift the amounts needed from final uses to domestically produce the avoided intermediate imports; (c) *total intermediate input coefficients applied abroad (common production technologies)*; the total (domestic and imported) input coefficients structure of the non-EU countries will be assumed to be the same as that of the EU countries.

Equations (7) and (8) are therefore simplified to the following expressions as long as these three above assumptions are made. Therefore:

$$cf_u = \hat{C}_u\left(I-\left(A^{uu}+A^{ru}\right)\right)^{-1}\left[y_{uu}+y_{ru}\right] \tag{9}$$

$$cf_r = \hat{C}_u\left(I-\left(A^{uu}+A^{ru}\right)\right)^{-1}\left[y_{ur}+y_{rr}\right] \tag{10}$$

Note that the information needed to make the calculations in (9) and (10) only refers to the EU region for which domestic and import data is assumed to be available. Table 7.1 summarizes the three different approaches that we will denote:[2] GEN for equations (5) and (6), DEM for equations (7) and (8) and TOT for equations (9) and (10).

Data sources and empirical results

The data sources refer to two main topics: emissions and economic data. The emission coefficients of the EU27 come from Eurostat whereas those for the non-EU countries were estimated by Neuwahl *et al.* (2010). The latter emission coefficients are preliminary and part of work that is currently being developed by the European Commission's Joint Research Centre (IPTS) for the seventh framework programme EU-funded project: World Input–Output Database: Construction and Applications. Technical details of the estimation of the environmental satellite accounts and their compilation can be provided to readers upon request to the author.

With regard to economic data, the sources come from two inter-related projects carried out at Eurostat, i.e. (a) the compilation of an annual time series of consolidated European supply–use tables and input–output tables for EU27 and the euro area (2000–2006), and (b) its application to the estimation of the

Table 7.1 Calculation of European/non-European carbon footprint

	Carbon footprint of EU countries	*Carbon footprint non-EU countries*
GEN	$\hat{C}_u L^{uu} y_{uu} + \hat{C}_u L^{ur} y_{ru} + \hat{C}_r L^{ru} y_{uu} + \hat{C}_r L^{rr} y_{ru}$	$\hat{C}_u L^{uu} y_{ur} + \hat{C}_u L^{ur} y_{rr} + \hat{C}_r L^{ru} y_{ur} + \hat{C}_r L^{rr} y_{rr}$
DEM	$\hat{C}_u[L^{uu} + L^{ru} y_{uu} + L^{ur} + L^{rr} y_{ru}]$	$\hat{C}_u[L^{uu} + L^{ru} y_{ur} + L^{ur} + L^{rr} y_{rr}]$
TOT	$\hat{C}_u\, I - A^{uu} + A^{ru\,-1}\, y_{uu} + y_{ru}$	$\hat{C}_u\, I - A^{uu} + A^{ru\,-1}\, y_{ur} + y_{rr}$

carbon footprint of the European domestic final demand and imports, separately. The former project has been conducted by Eurostat and the Joint Research Centre's IPTS of the European Commission (Rueda-Cantuche and Ritzmann, 2009 with support of the Constance University of Applied Science). The tables produced are supply tables at basic prices with a transformation to purchasers' prices, use tables at basic prices (broken down into uses of domestic production and imports) and symmetric input–output tables. This project will be the basis to a regular compilation of such tables by Eurostat. The estimation process is based on primary data sources transmitted by member states through the official ESA95 transmission programme and on additional (confidential) data provided by the National Statistical Institutes (NSI), e.g. valuation matrices and use tables at basic prices. The latter project was commissioned by Eurostat and performed by a consortium consisting of the Netherlands Organization for Applied Scientific Research (TNO), the Centre of Environmental Sciences of Leiden University (CML), the Norwegian University of Science and Technology (NTNU) and the University of Groningen (RUG); together with the additional support of the Joint Research Centre's IPTS of the European Commission. The outcomes of the two projects were made to be publicly available in May 2011.

However, the lack of available data on the input structures and the final demand of the non-EU countries led us to make the following assumptions for the most general approach (GEN): (a) split up the column vector of extra-EU exports into exports destined to intermediate uses and exports destined to final demand; we used the same shares of those of the consolidated domestic input–output tables of the EU27 for each year; (b) construct a fully fledged matrix of EU intermediate exports (A^{ur}) to the rest of the world from the column vector of total intermediate exports. The allocation was proportional to the row elements of the consolidated supply table of the EU27. This assumption implicitly involves that the supplier structure of commodities is the same in EU and non-EU countries, e.g. agricultural products are supplied to a large extent by the agricultural industry; (c) assume that the total input structures of the non-EU countries were the same as those of the EU countries (common production technology), i.e.: $A^{Tu} = A^{Tr}$. Therefore, if $A^{Tr} = A^{rr} + A^{ur}$, then $A^{rr} = A^{Tu} - A^{ur}$; (d) omit the calculation of the carbon footprint of the non-EU countries.

Provided a known estimated A^{ur} matrix and assuming a common production technology, we will proceed to introduce another assumption on top of these two, i.e. the use of common (EU) emission coefficients. The resulting nested set of assumptions corresponds to the so-called DEM approach (see previous section). Next, on top of the current three assumptions, we eventually assume changes in the shares of final and intermediate uses over the total commodity output of each region (see the Appendix for more details). The resulting set of nested assumptions corresponds now to the TOT approach in the previous section.

Table 7.2 shows a summary of the results obtained from the empirical calculations. Figures 7.1 to 7.4 will serve to illustrate the main findings of the results summarized in Table 7.2.

Table 7.2 Empirical results for calculation of European carbon footprint

Year	*Within EU (v_{uu})*		*Outside EU (v_{ru})*		*Carbon (cf_u)*
	Intra-regional	*Inter-regional*	*Intra-regional*	*Inter-regional*	*Total*
	Total embedded carbon in EU27 (GEN)				
2000	2,786,014	12,623	317,607	662,410	3,778,654
2001	2,817,339	13,694	324,303	686,332	3,841,668
2002	2,812,202	12,386	290,468	609,420	3,724,476
2003	2,902,688	9,513	235,868	492,104	3,640,174
2004	2,954,284	7,515	202,883	431,084	3,595,767
2005	2,924,772	7,617	220,261	486,222	3,638,871
2006	2,923,182	8,563	250,513	540,554	3,722,812
	Total embedded carbon in EU27 (DEM)				
2000	2,786,014	12,623	169,843	356,262	3,324,743
2001	2,817,339	13,694	165,482	349,959	3,346,474
2002	2,812,202	12,386	159,869	332,914	3,317,371
2003	2,902,688	9,513	168,288	351,876	3,432,365
2004	2,954,284	7,515	163,709	354,656	3,480,164
2005	2,924,772	7,617	176,203	393,574	3,502,165
2006	2,923,182	8,563	194,039	419,957	3,545,741
	Total embedded carbon in EU27 (TOT)				
2000	–	–	–	–	3,331,643
2001	–	–	–	–	3,371,551
2002	–	–	–	–	3,320,917
2003	–	–	–	–	3,437,387
2004	–	–	–	–	3,481,958
2005	–	–	–	–	3,535,546
2006	–	–	–	–	3,546,766

Source: own elaboration.

Note
Units: thousands of tonnes of CO_2.

Figure 7.1 clearly shows the differentiated shapes of the curves for the three different approaches. While the general approach (GEN) has a peak in 2001 and then falls down until 2004 to start growing again until 2006, the domestic emissions-based approach (DEM) and the total use-based approach (TOT) grow almost continuously during the whole period. Moreover, the difference in levels between the GEN and the DEM approaches is approximately 7.7 per cent lower for the latter. This result is extremely important since the estimated carbon footprint may be under-estimated if we do not use differentiated emission coefficients for EU and non-EU countries. In other words, since the rest of the world has apparently less clean technologies, an assumption of domestic emission coefficients will generally lower the real amount of pollution generated in those non-EU countries. The comparison between the GEN and the TOT approaches yields almost the same results (7.4 per cent) provided that the latter also assumes domestic emission coefficients but with somewhat different input structures.

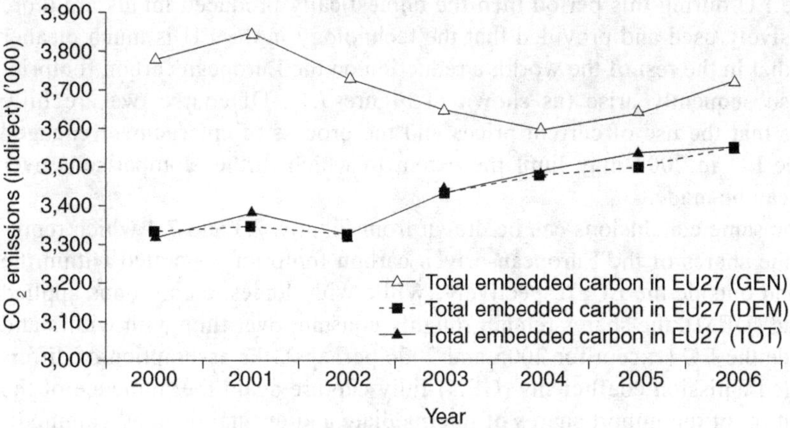

Figure 7.1 European carbon footprint: comparison of various methodologies (2000–2006).

Nevertheless, the shapes of the curves are resoundingly different and we wonder why this is. So far, we have an answer that justifies the differences in levels but not the shape of the curves. To this end, Figure 7.2 casts some light on the issue. During the period 2000–2006, the import shares over the total industry outputs decreased permanently until 2004 and then started to grow again. Similar but smoother behaviour can be seen in the import shares of final demand of commodities. Interestingly, the evolution of the import shares closely resembles the GEN carbon footprint approach (see Figure 7.1).

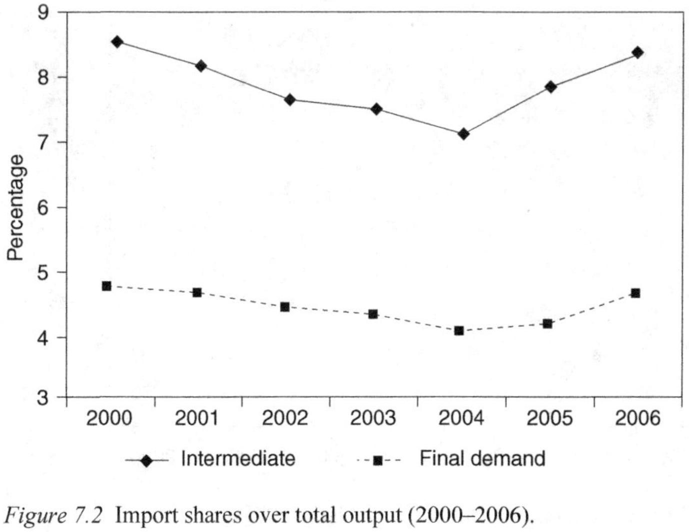

Figure 7.2 Import shares over total output (2000–2006).

Indeed, if import shares of intermediate and/or final demand inputs decrease in the EU during this period then the domestically produced inputs are more intensively used and provided that the technology in the EU is much cleaner than that in the rest of the world, a reduction on the European carbon footprint will subsequently arise (as shown in Figure 7.1). Of course, we are fully aware that the use of current prices and the process of enlargement engaged by the EU in 2004 may limit the extent to which further comparisons over time can be made.

The same conclusions can be drawn from Figures 7.3 and 7.4 which represent the shares of the European-driven carbon footprint generated within the EU and outside the EU, respectively. While with domestic emissions applied abroad (DEM), the shares remain roughly constant over time both within and outside the EU (except for 2005 and 2006 perhaps), the assumption of differentiated emission coefficients (GEN) fully capture again the influence of the evolution of the import shares of intermediate and/or final demand commodities on the shares of the European carbon footprint generated outside the EU countries (see Figure 7.4). Evidently, a decrease in the import shares of inputs also leads to reductions in the shares of the European carbon footprint emitted in the rest of the world. Also interestingly, the shape of the curve of the part of the European carbon footprint generated within the EU countries (see Figure 7.3) closely resembles the evolution of the overall European carbon footprint when equalizing domestic emissions across all countries (DEM).

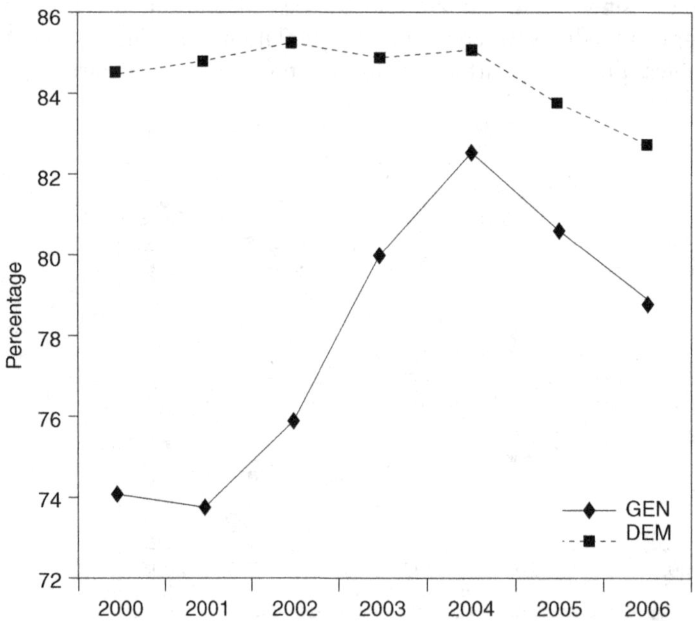

Figure 7.3 EU carbon footprint within the EU (2000–2006).

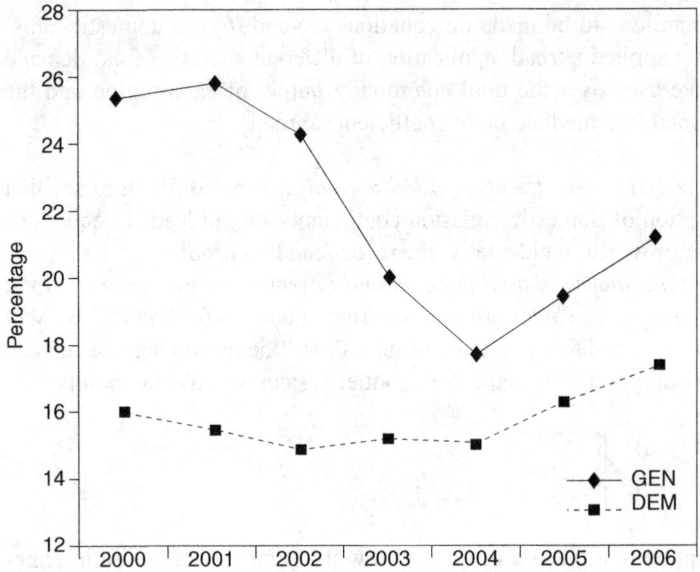

Figure 7.4 EU carbon footprint outside the EU (2000–2006).

Conclusions

In this chapter, we have set up a theoretical multi-regional framework (of two regions, Europe and rest of the world) with which the European carbon footprint can be calculated under the ideal conditions of full information availability. Step by step, we have introduced assumptions due to restrictions in the data so that three different nested approaches are identified and discussed, namely: (a) common production technology with different emission coefficients by region; (b) common production technology with common domestic emission coefficients for EU and non-EU countries; and (c) common production technology, common (EU) emission coefficients plus some additional assumptions on the inter-regional shares between intermediate and final demand of commodities. The latter approach turned out to be the result of applying the Leontief inverse to the matrix of total (domestic and imported) technical coefficients. The results have shown that the assumption of domestic emission coefficients has serious consequences on the estimation of the European carbon footprint mainly due to the fact that it does not capture the changes in the import shares of intermediate and/or final uses.

Therefore, our recommendation for future carbon footprint analyses is to make additional efforts in estimating differentiated emission coefficients for the regions under study and limit the use of domestic emissions abroad as much as possible. The results of the carbon footprint if other methods are adopted seem to be seriously distorted and largely under-estimated.

Appendix

The three assumptions to be made on equations (5) and (6) are domestic emission coefficients applied abroad, application of different shares of final demand and intermediate uses over the total commodity output of each region and the application of total intermediate input coefficients abroad.

a *Domestic emission coefficients applied abroad*; it is not difficult to see that the assumption of domestic emission coefficients abroad leads to equations (4) and (5), of which, incidentally, the reader can find proof.

b *Changes in the shares of final demand and intermediate uses over the total commodity output of each region*; next, from equations (4) and (5), we will derive a more simplified expression that will still be used when we refer to the third assumption. If we take the Leontief system as a starting point:

$$\begin{bmatrix} x_u \\ x_r \end{bmatrix} = \begin{bmatrix} A^{uu} & A^{ur} \\ A^{ru} & A^{rr} \end{bmatrix} \begin{bmatrix} x_u \\ x_r \end{bmatrix} + \begin{bmatrix} y_{uu} + y_{ur} \\ y_{ru} + y_{rr} \end{bmatrix} \tag{A.1}$$

the assumption that the rest of the world will not import any more intermediate inputs to the EU but only commodities for final uses implies a shift from the top-right element of the partitioned A-matrix to the top element of the final demand vector. Similarly, the formerly imported intermediate inputs must be produced now domestically by the rest of the world, thus causing a new shift from the bottom element of the final demand vector to the bottom-right element of the A-matrix in order to satisfy this new demand for domestically (of the rest of the world) produced intermediate inputs. Mathematically speaking, the Leontief system becomes:

$$\begin{aligned} \begin{bmatrix} x_u \\ x_r \end{bmatrix} &= \begin{bmatrix} A^{uu} & A^{ur} - A^{ur} \\ A^{ru} & A^{rr} + A^{ur} \end{bmatrix} \begin{bmatrix} x_u \\ x_r \end{bmatrix} + \begin{bmatrix} y_{uu} + \left(y_{ur} + A^{ur} x_r \right) \\ y_{ru} + \left(y_{rr} - A^{ur} x_r \right) \end{bmatrix} \\ &= \begin{bmatrix} A^{uu} & 0 \\ A^{ru} & A^{Tr} \end{bmatrix} \begin{bmatrix} x_u \\ x_r \end{bmatrix} + \begin{bmatrix} y_{uu} + \left(y_{ur} + A^{ur} x_r \right) \\ y_{ru} + \left(y_{rr} - A^{ur} x_r \right) \end{bmatrix} \end{aligned} \tag{A.2}$$

with A^{Tr} being the total input coefficient matrix of the rest of the world (non-EU countries). Note that the elements included in parentheses at the top of the final demand vector make the total exports of the EU. This has been recognized as a big advantage of this assumption since it helps in the calculation of the carbon footprint provided that information on EU exports is widely accessible. However, there is still uncertainty on the total input coefficients of the rest of the world (A^{Tr}).

c *Total intermediate input coefficients applied abroad (common production technology)*; taking (A.2) as a starting point, we will now assume that the total input structure of the non-EU countries is exactly the same as that of the EU countries which is the same as: $A^{Tr} = A^{uu} + A^{ru} = A^{Tu}$. (A.2) can therefore be re-written as:

$$\begin{bmatrix} x_u \\ x_r \end{bmatrix} = \begin{bmatrix} A^{uu} & 0 \\ A^{ru} & A^{Tu} \end{bmatrix} \begin{bmatrix} x_u \\ x_r \end{bmatrix} + \begin{bmatrix} y_{uu} + (y_{ur} + A^{ur} x_r) \\ y_{ru} + (y_{rr} - A^{ur} x_r) \end{bmatrix}$$

and subsequently,

$$\begin{bmatrix} x_u \\ x_r \end{bmatrix} = \left(\begin{bmatrix} I & 0 \\ 0 & I \end{bmatrix} - \begin{bmatrix} A^{uu} & 0 \\ A^{ru} & A^{Tu} \end{bmatrix} \right)^{-1} \begin{bmatrix} y_{uu} + (y_{ur} + A^{ur} x_r) \\ y_{ru} + (y_{rr} - A^{ur} x_r) \end{bmatrix}$$

$$\begin{bmatrix} x_u \\ x_r \end{bmatrix} = \begin{pmatrix} I - A^{uu} & 0 \\ -A^{ru} & I - A^{Tu} \end{pmatrix}^{-1} \begin{bmatrix} y_{uu} + (y_{ur} + A^{ur} x_r) \\ y_{ru} + (y_{rr} - A^{ur} x_r) \end{bmatrix}$$

$$\begin{bmatrix} x_u \\ x_r \end{bmatrix} = \begin{pmatrix} L^{uu} & 0 \\ L^{ru} & L^{rr} \end{pmatrix} \begin{bmatrix} y_{uu} + (y_{ur} + A^{ur} x_r) \\ y_{ru} + (y_{rr} - A^{ur} x_r) \end{bmatrix} =$$

$$\begin{pmatrix} (I - A^{uu})^{-1} & 0 \\ (I - A^{Tu})^{-1} A^{ru} (I - A^{uu})^{-1} & (I - A^{Tu})^{-1} \end{pmatrix} \begin{bmatrix} y_{uu} + (y_{ur} + A^{ur} x_r) \\ y_{ru} + (y_{rr} - A^{ur} x_r) \end{bmatrix} \quad \text{(A.3)}$$

Next, let us apply our previous equations (7) and (8) to this new Leontief inverse matrix. For the EU carbon footprint, (7) yields:

$$\begin{aligned} cf_u &= \hat{C}_u \left[(L^{uu} + L^{ru}) y_{uu} + (L^{ur} + L^{rr}) y_{ru} \right] = \hat{C}_u \left[(L^{uu} + L^{ru}) y_{uu} + (0 + L^{rr}) y_{ru} \right] \\ &= \hat{C}_u \left[(L^{uu} + L^{ru}) y_{uu} + (I - A^{Tu})^{-1} y_{ru} \right] \end{aligned}$$

and for the non-EU countries:

$$\begin{aligned} cf_r &= \hat{C}_u \left[(L^{uu} + L^{ru})(y_{ur} + A^{ur} x_r) + (L^{ur} + L^{rr})(y_{rr} - A^{ur} x_r) \right] \\ &= \hat{C}_u \left[(L^{uu} + L^{ru})(y_{ur} + A^{ur} x_r) + (0 + L^{rr})(y_{rr} - A^{ur} x_r) \right] \\ &= \hat{C}_u \left[(L^{uu} + L^{ru})(y_{ur} + A^{ur} x_r) + (I - A^{Tu})^{-1} (y_{rr} - A^{ur} x_r) \right] \end{aligned}$$

The last task to finalize the proof is to find out what $(L^{uu} + L^{ru})$ is in terms of the A-matrices. From (A.3),

$$L^{uu} + L^{ru} = (I - A^{uu})^{-1} + (I - A^{Tu})^{-1} A^{ru} (I - A^{uu})^{-1}$$

and by pre-multiplying both sides by $(I - A^{Tu})^{-1}$,

$$
\begin{aligned}
&(I - A^{Tu})(L^{uu} + L^{ru}) \\
&\quad = (I - A^{Tu})(I - A^{uu})^{-1} + (I - A^{Tu})(I - A^{Tu})^{-1} A^{ru}(I - A^{uu})^{-1} \\
&\quad = (I - A^{Tu})(I - A^{uu})^{-1} + A^{ru}(I - A^{uu})^{-1} \\
&\quad = (I - A^{uu})^{-1} - A^{Tu}(I - A^{uu})^{-1} + A^{ru}(I - A^{uu})^{-1} \\
&\quad = (I - A^{uu})^{-1} - (A^{uu} + A^{ru})(I - A^{uu})^{-1} + A^{ru}(I - A^{uu})^{-1} \\
&\quad = (I - A^{uu})^{-1} - (A^{uu})(I - A^{uu})^{-1} - A^{ru}(I - A^{uu})^{-1} + A^{ru}(I - A^{uu})^{-1} \\
&\quad = (I - A^{uu})^{-1} - (A^{uu})(I - A^{uu})^{-1} = (I - A^{uu})(I - A^{uu})^{-1} = I
\end{aligned}
$$

which means that,

$$(L^{uu} + L^{ru}) = (I - A^{Tu})^{-1}$$

and for the EU carbon footprint,

$$
\begin{aligned}
cf_u &= \hat{C}_u\left((L^{uu} + L^{ru})y_{uu} + (I - A^{Tu})^{-1} y_{ru}\right) \\
&= \hat{C}_u\left((I - A^{Tu})^{-1} y_{uu} + (I - A^{Tu})^{-1} y_{ru}\right) \\
&= \hat{C}_u (I - A^{Tu})^{-1}[y_{uu} + y_{ru}] = \hat{C}_u (I - A^{Tu})^{-1}[y_{Tu}]
\end{aligned}
$$

and for the non-EU countries,

$$
\begin{aligned}
cf_r &= \hat{C}_u\left[(L^{uu} + L^{ru})(y_{ur} + A^{ur}x_r) + (L^{ur} + L^{rr})(y_{rr} - A^{ur}x_r)\right] \\
&= \hat{C}_u\left[(L^{uu} + L^{ru})(y_{ur} + A^{ur}x_r) + (0 + L^{rr})(y_{rr} - A^{ur}x_r)\right] \\
&= \hat{C}_u\left[(I - A^{Tu})^{-1}(y_{ur} + A^{ur}x_r) + (I - A^{Tu})^{-1}(y_{rr} - A^{ur}x_r)\right] \\
&= \hat{C}_u\left[(I - A^{Tu})^{-1}(y_{ur} + A^{ur}x_r + y_{rr} - A^{ur}x_r)\right] \\
&= \hat{C}_u\left[(I - A^{Tu})^{-1} y_{Tr}\right]
\end{aligned}
$$

Acknowledgements

The author gratefully acknowledges the valuable comments made by Erik Dietzenbacher (University of Groningen), Stephan Moll (Eurostat), Iñaki Arto (Joint Research Centre – IPTS) and Richard Wood (NTNU) on previous versions of this work, the provision of the environmental data for the non-EU countries by Frederik Neuwahl, Aurelian Genty and Alejandro Villanueva (Joint Research Centre – IPTS), the methodological support and/or the compilation of the national and consolidated supply–use and input–output tables for the EU27 provided by Joerg Beutel (Konstanz University of Applied Sciences), Isabelle Remond-Tiédrez (Eurostat), Anne Foltete (Eurostat), Gerard Hanney (Eurostat), Maaike Bouwmeester and Jan Oosterhaven (University of Groningen) and the calculations of the European carbon footprint for the TOT approach provided by Richard Wood (NTNU) and Arnold Tukker (TNO) for Eurostat. The author especially wishes to thank his late colleague P. Ritzmann without whom this work would not have been possible. The views expressed in this chapter are those of the author and should not be attributed to the European Commission or its services.

Notes

1 Matrices are indicated by italicized upright capital letters; vectors by italicized lowercase letters. Vectors are columns by definition and row vectors are defined by transposition, indicated by a prime. Matrices with all the off-diagonal elements set to zero and the elements of a vector placed on the main diagonal are denoted by a circumflex.

2 GEN stands for 'General' since it refers to the general case, DEM refers to 'Domestic Emissions' since they are assumed to be domestic emissions abroad and TOT comes from 'Total' due to the fact that the total input coefficient matrix is used.

References

Dietzenbacher, E. (2002) 'Interregional multipliers: looking backward, looking forward', *Regional Studies*, 36: 125–136.

Erumban, A.A., Gouma, F.R., Los, B., Stehrer, R., Temurshoev, U., Timmer, M.P. and de Vries, G.J. (2010) 'World input–output database: construction, challenges and applications', paper presented at the 31st General Conference of the International Association for Research in Income and Wealth, Sankt-Gallen, Switzerland.

Neuwahl, F., Genty, A. and Villanueva, A. (2010) 'Preliminary database of environmental satellite accounts: technical report on their compilation', WIOD Deliverable 4.2 (internal documentation), WIOD Project. Online: www.wiod.org.

Rueda-Cantuche, J.M. and Ritzmann, P. (2009) 'A consolidated European Union and euro area supply-use system and input–output tables', paper presented at the 17th International Input–Output Conference, São Paulo, Brazil.

Serrano, M. and Dietzenbacher, E. (2010) 'Responsibility and trade emission balances: an evaluation of approaches', *Ecological Economics*, 69: 2224–2232.

Tukker, A., *et al.* (2009) 'Towards a global multi-regional environmentally extended input–output database', *Ecological Economics*, 68: 1928–1937.

8 Aggregation bias in 'consumption vs. production perspective' comparisons

Evidence using the Italian and Spanish NAMEAs

Giovanni Marin, Massimiliano Mazzanti and Anna Montini

Introduction

The integration of the National Accounting Matrix including Environmental Accounts (NAMEA) and input–output (I–O) tables (usually referred to as environmentally extended input–output analysis – EE-IOA based on NAMEA data) is a challenging but promising way to analyse the factors behind income–environment relationships in international settings (Cole, 2004; Copeland and Taylor, 2004; Frankel and Rose, 2005). More specifically, it can be used to disentangle production and consumption perspectives through the detailed sector-based information provided by the two frameworks. National and international sources of environmental effects can be ascertained in strict connection with streams of literature such as the 'ecological footprint' kind of analysis and decomposition analyses. It is also heavily embedded in the wide realm that deals with sustainable consumption and production (SCP) issues (Harris, 2001), a key pillar of current and future EU policy efforts.

A comparison of the production vs. consumption perspective can have important policy implications. Traditionally, environmental policy focused mainly on production activities as sources of impacts and the actors to be targeted by legislation and regulation. Looking at the role of final consumption for vertically integrated domestic and international impacts can push policy attention towards the possible role of the consumer as an actor of environmental policies, together with the international responsibility for spillover of impacts abroad. A key issue here relates to the modelling of the technology associated with imported goods (produced abroad by the stimulus of domestic consumption), which is tricky in practice given the scarcity of data at that level of detail and at sector level. Given the technology, net trade-embodied pollution arises as a structural phenomenon from a systematic difference in the composition of domestic production compared with the composition of consumption rather than structural level imbalances than cannot be sustainable in time. Systematic differences in turn arise from a production specialization that can be, and usually is, more marked than the consumption specialization of a country.

Current consumption structure is changing slowly in most European countries, although with significant momentum, whereas production specialization is changing faster. In a dynamic setting, consumer behaviour is changing too slowly in terms of embodied environmental efficiency compared with domestic production, thus possibly creating a net demand of pollution abroad through net import. Although consumption structure and behaviour can be less sensitive to environmental policies than production, e.g. due to lacking legal basis to constrain the freedom of choice, there can be room for addressing consumers and their behaviour to contribute to higher efficiency in terms of vertically integrated environmental impacts. The EU strategies on SCP paves the way to this policy direction, and analyses based on EE-IOA, addressing the differences between the two perspectives, can clarify the needs and implications of these policies.

We can affirm that sector-based input–output datasets existing for EU countries offer the possibility of highlighting how emissions are indirectly associated with production. NAMEA-type tables are datasets with coefficients on emission per output that can thus be matched with I–O tables for useful integration. Integration aims at calculating economic–environmental performances by sector by including the role of trade. In other words, it aims to test the hypothesis that given different relative emission efficiency, the structure of imports and exports matters. EE-IOA can provide additional information on environmental implications of economic structure and structural change; its objective is to investigate to what extent changes in final consumption patterns, production technologies and trade patterns (as a result of the decoupling of consumption from production) affect domestic and world-induced air emissions. Moreover, EE-IOA quantifies to what extent the geographical separation between consumption and production activities has occurred and whether it has determined increases or decreases in global environmental pressures.

From a general and methodological point of view, the integration of NAMEA accounting and I–O tables touches upon ecological/environmental economics and industrial ecology frameworks. Due to the striking increase of related works in such realms, the brief survey we provide in the next paragraph aims to give insights into recent developments and offer stimulus for future analyses rather than offering full coverage. It is worth noting that, very recently, there has been increasing interest in these environmental issues in the I–O society. Plenty of environmental papers appeared in the 2010 Sidney Conference of the I–O Society, some of those exploiting NAMEA. A boom of papers on environmentally extended I–O was reached in 2009 that witnessed a peak (Hoekstra, 2010), with a total of 360 papers, from 1969 to 2010. A related field of analyses which has witnessed great relevance in the I–O arena is structural decomposition analysis (SDA), a core technique for analysing factors behind delinking that focuses on sector heterogeneity. Decomposition analysis is one of the most effective and widely applied tools for investigating the mechanisms influencing energy consumption and emissions and their environmental side-effects (Mazzanti and

Montini, 2010). SDA has been applied to a wide range of topics, including demand for energy (e.g. Jacobsen, 2000; Kagawa and Inamura, 2004, 2001) and pollutant emissions (e.g. Casler and Rose, 1998; Wier, 1998). Many studies address industry. Nevertheless, services are also relevant: they are less energy intensive but present lower technological contents and can indirectly contribute to strong environmental impacts (we note the NAMEA-based disentangled analyses in Marin and Mazzanti (in press), who present industry vs. services assessments for Italy). Alcántara and Padilla (2009) analyse CO_2 emissions for Spain using I–O (year 2000). They conclude that

> Transport activities are the services with the highest level of the direct emissions generated in the production of the sector. These activities are required by the other sectors of the economy to a greater degree than they are for their own final demand. Therefore, the production sold to other sectors causes more emissions than its own final demand. However, in the case of other service activities, direct and indirect emissions related to final demand are much more important, due to the strong pull effect of service activities on other activities of the economy. In this respect, Wholesale and retail trade, Hotels and restaurants, Real estate, renting and business activities, and Public administration services should be highlighted. These services receive scarce attention in the design of policies aimed at reducing emissions, but are notably responsible for the major increase in emissions experienced in recent years.

Trade is the key factor in recent extended I–O and NAMEA works that aim to deal with SCP contents.[1] We recall that the main aim is to assess direct and indirect environmental effects by attributing their relative weights to national consumption and to exports in the explanation of a country's environmental performance. We present below a brief discussion which centres on recent and comparable works that have touched the core issue of I–O/NAMEA integration with the aim of analysing the sustainability of production and consumption by taking trade into account. Currently, main efforts aim to move away from the domestic technology assumption (DTA) that says that imported goods use the same technology (in terms of structure of intermediate inputs and environmental efficiency) as goods produced domestically.

A very recent example is Arto *et al.* (2010). They show that Spain is a net emissions exporter and consequently, its consumer responsibility in emissions is higher than its producer responsibility. The difference between both types of responsibility increases by applying the physical DTA. This is substantially due to the fact that the monetary DTA estimates less embodied emissions in imports from non-Annex I countries than the physical DTA.[2] Linking to NAMEA studies commented on below, we note Watson and Moll (2008). It can be seen that the NAMEAs can be manipulated using EE-IOA to provide the two different perspectives useful for SCP: usual production perspective and an additional consumption perspective. The consumption perspective focuses on the production chains of all final products consumed domestically. This includes goods produced in the home country for domestic consumption and products imported for consumption, but it excludes production chains of exports. We also highlight the recent special issue in *Economic Systems Research*, (Volume 21, Issue 4, 2009) that deals with applications regarding I–O and carbon footprinting. A study that brings together various frameworks highlighting flexibility of methods and usefulness of integrated use is certainly Moll *et al.* (2007) (two years, eight countries). The work shows that, according to different sectors and countries, the domestic production patterns and associated direct domestic environmental pressures are rather different. Electricity, gas and hot water production, agriculture and transport and communication services cause the majority of environmental pressures. Direct pressures from private households (mainly for heating and private transport) constitute another important source.

With regard to international factors, it can be seen that a second determinant for cross-country differences in domestic direct pressures is the role of exports. When it comes to consumption and investment patterns, Moll *et al.* (2007) show that cross-country differences are far less pronounced than production patterns. Analyses focusing on environmental impacts of consumption (by categories) are also found in Huppes *et al.* (2005): food, heating and transport emerge as core impacting aggregation.[3] We also note the extensive IPTS 'EIPRO' report (2006). In general, it is the satisfaction and organization of basic needs, i.e. eating, housing and mobility, that is responsible for the majority of production-cycle-wide environmental pressures.

Finally, we cited the peak of emphasis that the I–O society gave to environmentally extended I–O analyses in 2010. Among the many works presented, in our view, some of the more intriguing pieces that tried to exploit NAMEA accounting in I–O contexts were by Rueda-Cantuche (2010), Hoekstra *et al.* (2010), Basina and Sfetsos (2010), Rormose *et al.* (2010). Three of the authors mentioned appear in this book with some complementary pieces of work.

Methodological and empirical literature

Empirical analysis with an extension of the use of the statistical information derived from environmental accounts and the input–output tables requires several considerations to be made. The aim of this chapter is linked to the so-called *aggregation bias*. As suggested by Lenzen (2011), environmental I–O analyses of environmental issues are often plagued by the fact that environmental and I–O data exist in different classifications.

A recurring problem in EE-IOA is that input–output accounts and environmental statistics are often not compiled by the same statistical agency and therefore often differ with respect to the classification of economic sectors and other definitions. In these cases, analysts have to carry out data collection and harmonization procedures in order to integrate both accounts. What can happen is that: (a) environmentally sensitive sectors are sometimes more aggregated in the economic I–O database than the environmental dataset because monetary I–O tables are compiled with no environmental implications in mind; (b) I–O data are disaggregated into more sectors than environmental satellite data, especially for services sectors (Lenzen, 2011).

There are two basic alternatives for dealing with such a misalignment: either environmental data have to be aggregated into the I–O classification (but some environmental sensitive data will lose their peculiarities) or I–O data have to be disaggregated based on fragmentary information (with several assumptions).

By keeping this in mind, the aggregation bias is likely to severely affect the construction of environmentally extended Multi-Region Input–Output (EE-MRIO) analysis, as recently suggested by Su *et al.* (2010) and Lenzen (2011), as well as environmentally extended Single Region Input–Output accounts with specific assumption regarding the technology used (embodied in international trade, specifically those in the import data).

As will be explained below, the DTA relies on the consideration that all imported commodities are produced with the same mix of intermediate inputs (in monetary terms and as indicated by the intermediate flows in the input–output table) and with the same environmental efficiency (in terms of emissions per monetary unit of output) as domestic commodities.

Some authors (including Turner *et al.*, 2007; Peters, 2007; Serrano and Dietzenbacher, 2010; Arto *et al.*, 2010) suggest moving away from the DTA because they consider it too simplistic but they recognize that, generally, the DTA produces better estimates than ignoring imports altogether. Ideally, full information on bilateral trade plus corresponding NAMEA data by country is equivalent to analysing trade of impacts at country-by-country differentiated coefficients. However, it requires a wide and often unavailable range of data. A possibility for dealing with the latter is to include only the most important trade partner (and for which required data are available) in terms of emissions embodied in imports and this, as suggested by Andrew *et al.* (2009), can substantially improve the accuracy of estimates. For the emissions embodied in imports, Andrew *et al.* (2009) find that the unidirectional trade model gives a good approximation to the full MRIO model when the number of regions in the model is small. Moreover, the assumption that imports are produced with DTA in an MRIO model can introduce significant errors and requires careful validation before results are used.

If we re-examine the issue of aggregation bias, the studies that have analysed the CO_2 emissions embodied in international trade have also been carried out by using an input–output framework at a specific level of sector aggregation. Generally, the choice has been made to a large extent according to economic and energy data availability or, similarly, economic and environmental data availability. A finding for Su *et al.* (2010) is that levels of around 40 sectors appear to be sufficient to capture the overall share of emissions embodied in a country's exports.

The issues related to aggregation bias and a possible DTA obviously affect the consumption[4] and production perspective when looking at the corresponding emissions. As suggested in the introduction, the focus of the EU policy area on

SCP forces researchers to consider new tools of analysis and one of them is the EE-IOA based on NAMEA data. The notion of responsibility (either for the consumer or the producer) addresses some considerations.

As suggested by Gallego and Lenzen (2005), there is a sort of domination of producer-centric representation to view the environmental or social impacts of industrial production. When thinking about environmental impacts, crucial questions arise such as who is responsible for what? How is the responsibility to be shared? Should a firm have to improve the eco-friendliness of its products, or it is up to the consumer to buy or not to buy? Questions of this type can be considered when deciding who takes the credit for successful abatement measures that involve producers and consumers. Moreover, the kind of pollutant considered influences policy implications when looking at the ratio between consumption-based emissions (C) and producer-based emissions (P). If we consider global pollutants, such as CO_2, and C is bigger than P, the country responsibility is bigger than that reported by the official statistics. If we consider local pollutants and C is bigger than P, the country would be displacing environmental costs to other territories.

Gallego and Lenzen (2005) propose a method of re-tracing the flow of past inter-industrial transactions to allocate responsibility for production impacts consistently among all agents such as consumer, producers, workers and investors. According to them, the input–output analysis can be used as a descriptive tool to re-trace the flow of past transactions and examine ex post how, for example, inputs of resources or outputs of pollution were associated with these transactions.

Serrano and Dietzenbacher (2010) define two ways to evaluate the international responsibility of emissions generated by one country – in their analysis they consider Spain in 1995 and 2000 and nine gases – that were shown to be equivalent: the trade emission balance (as the difference between the emissions embodied in a country's exports and imports) and the responsibility emission balance (as the difference between the responsibility of one country as a producer and its responsibility as a 'consumer').

Methodology and data

In this section, we outline the main features of the domestic technology assumption (DTA henceforth) and we summarize the main issues related to the assessment of the aggregation bias in input–output analysis including NAMEA data.

Domestic technology assumption and aggregation bias

The domestic technology assumption

The hypothesis behind the DTA is that the imported commodities (either as intermediate inputs or final consumption) are produced with the same mix of intermediate inputs (in monetary terms) and with the same environmental efficiency (in terms of emissions per monetary unit of output) as domestic commodities.

Serrano and Dietzenbacher (2010) formally describe how and under which conditions an environmentally extended MRIO model accounting for worldwide induced emissions could be reduced to a model using only domestic data with an explicit DTA. In addition to assumptions on technology (i.e. the structure of intermediate inputs described by the input–output matrix) and on the vector of emission coefficients, the export of the country on which the analysis is focused should represent a negligible share of world output.

Another requirement, related to the validity of the domestic technology as a proxy of world technology, is that the country produces domestically at least part of all the commodities it consumes as intermediate inputs or final products. For example, this requirement is not fulfilled when a country has no particular raw materials in its soil or subsoil (oil, coal, gas, minerals, metals, etc.) and it is completely dependent on importing these commodities. As a result, the technology for the extracting industries (section C of NACE 1.1) in the input–output tables is biased towards secondary activities within the sector (e.g. basic transformation of raw materials) and it does not describe the main activity (i.e. extraction) properly. This problem is particularly relevant in environmentally extended input–output analyses in which extracting sectors are, in general, among the most polluting industries.

Although the DTA cannot be used to interpret the results as 'actual worldwide emissions induced by domestic final demand', it gives information on the potential emissions arising because of domestic final demand if the country has produced domestically the necessary final and intermediate goods (that is, using domestic technology). Estimates using the DTA, if interpreted properly, are therefore a particularly important indicator of consumer responsibility because of its low requirement for data, the possibility of replicating its results and the straightforward and clear hypothesis behind its implementation. For this reason, we claim that estimates based on the DTA should be used as a benchmark in more complex multi-regional environmentally extended input–output analysis aimed at assessing consumer responsibility.

However, the DTA and the overall EE-IOA results might be severely biased when the commodity/sector aggregation is very low and/or when the country which is analysed relies exclusively on import for certain commodities. In the latter case, in fact, either it will not possible to compute any domestic environmental coefficient (because both emissions and output are zero) or, if this sector is aggregated with other sectors, both the technology (the row of the matrix of technical coefficients when considering both imported and domestic intermediate inputs) and the emission coefficient of the aggregated sector could fail to represent technically viable technologies. A possible solution to this problem, although not conclusive, would be to substitute the specific rows of the matrix of technical coefficients and the specific entries of the vector of emission coefficient for these sectors with data of similar countries which have domestic production in these sectors. However, on the one hand, this kind of manipulation is likely to unbalance the whole input–output system and, on the other, the similarity is difficult to check due to the variety of dimensions included in this type of EE-IOA.

Before discussing the way in which aggregation is likely to introduce biases in the estimates of the level of emissions induced by final domestic demand, we will introduce some notation and explain how induced emissions are computed. The notation is summarized in Table 8.1.

When estimating the emissions induced worldwide by domestic final demand, we need to account for the intermediate inputs induced worldwide (thus using $\mathbf{L}_{d+m}$ as Leontief inverse) and for domestic final demand only ($\mathbf{f}_d$).

Induced emissions (consumption perspective, $\mathbf{e}_{cp}$) classified by product/industry are given by:

$$\mathbf{e}_{cp} = \left(\mathbf{b}'\,\mathbf{L}_{d+m}\left\langle \mathbf{f}_d \right\rangle\right)' \tag{1}$$

while total induced emissions ($\mathbf{e}_{cp}^{tot}$) may be obtained by post-multiplying $\mathbf{e}_{cp}$ by $\mathbf{i}$.[5]

Aggregation bias

The issue of the choice of the level of aggregation is crucial in any empirical analysis in economics.[6] Each aggregation results in losses of relevant information

Table 8.1 Summary of the relevant notation

Symbol	*Dimension*	*Description*
$\mathbf{Z}_d$	$n \times n$	Matrix of domestic intermediate inputs
$\mathbf{Z}_m$	$n \times n$	Matrix of imported intermediate inputs
$\mathbf{f}_d^d$	$n \times 1$	Vector of domestic final demand for goods produced domestically
$\mathbf{f}_d^m$	$n \times 1$	Vector of domestic final demand for goods produced in foreign countries (import of final goods)
$\mathbf{f}_x^d$	$n \times 1$	Vector of foreign final demand for goods produced domestically (export of final goods)
$\mathbf{f}_x^m$	$n \times 1$	Vector of foreign final demand for goods produced in foreign countries (re-export)
$\mathbf{e}$	$n \times 1$	Vector of domestic direct air emissions
$\mathbf{i}$	$n \times 1$	Summation vector (column vector of 1s)
$\mathbf{I}$	$n \times n$	Identity matrix
$\mathbf{S}$	$m \times n$	Aggregation matrix
$\mathbf{x}_d$	$n \times 1$	Domestic output ($\mathbf{Z}_d\mathbf{i}+\mathbf{f}_d^{\,d}+\mathbf{f}_x^{\,d}$)
$\mathbf{x}_{d+m}$	$n \times 1$	Domestic+imported output ($\mathbf{x}_d+\mathbf{Z}_m\mathbf{i}+\mathbf{f}_d^{\,m}+\mathbf{f}_x^{\,m}$)
$\mathbf{A}_{d+m}$	$n \times n$	Matrix of technical coefficients under the domestic technology assumption ($[\mathbf{Z}_d+\mathbf{Z}_m]<\mathbf{x}_{d+m}>^{-1}$)*
$\mathbf{L}_{d+m}$	$n \times n$	Leontief inverse under the domestic technology assumption $(\mathbf{I}-\mathbf{A}_{d+m})^{-1}$
$\mathbf{f}_d$	$n \times 1$	Domestic final demand ($\mathbf{f}_d^{\,d}+\mathbf{f}_d^{\,m}$)
$\mathbf{b}$	$n \times 1$	Emission coefficients ($\mathbf{e}<\mathbf{x}_d>^{-1}$)

Note

* <r> refers to a diagonal matrix with the diagonal composed by the elements of the vector **r**.

and in implicit compensations which are likely to affect the reliability of the results of any empirical analysis. However, aggregation is often unavoidable. First, the most common constraint regards the availability of sufficiently disaggregated raw data. Second, privacy legislation often prevents the diffusion of disaggregated data.[7] Third, time and computation constraints are likely to induce the researcher to employ readily available and small bases of aggregated data. Finally, when matching various sources of raw data, there is little alternative to aggregation if one or more of the sources is not sufficiently disaggregated, leading to an overall aggregation. This last issue is very common in multi-regional input–output models and the general approach involves reducing the overall level of disaggregation to the level of the most aggregated country/region.[8]

In environmentally extended input–output analysis, aggregation consists of a reduction in n sectors due to data availability constraints. More generally, if either the intermediate input matrices ($\mathbf{Z}_d$ or $\mathbf{Z}_m$) or the vector of direct emissions ($\mathbf{e}$) has low disaggregation, it is enough to force the researcher to reduce the level of aggregation of the model to the lowest 'n' dimension. This problem is particularly important when using MRIO models because in principle it is sufficient to have data limitations in one dimension of one out of several countries to force the researcher to reduce the overall level of disaggregation.

More formally, the way in which we estimate embodied emissions under different aggregations ($\mathbf{e}_{cp}^{aggr}$) is described by equation (2):

$$\mathbf{e}_{cp}^{aggr} = (\mathbf{e}'\mathbf{S}' <\mathbf{S}\,\mathbf{x}_d>^{-1})\,(\mathbf{I}\text{-}\ \mathbf{S}\,\mathbf{Z}_{d+m}\,\mathbf{S}' <\mathbf{S}\,\mathbf{x}_d>^{-1})^{-1}\,\mathbf{S} <\mathbf{f}_d>\mathbf{S}')' \neq \mathbf{S}\,\mathbf{e}_{cp} \quad (2)$$

where $\mathbf{S}$ is the aggregation matrix. An aggregation matrix is a rectangular matrix (in our case $m \times n$, with $m<n$) composed by 1s and 0s. The column sum of $\mathbf{S}$ will be 1 for each column while the sum of all the entries equals n. Pre-multiplying a column vector by $\mathbf{S}$ results in a new vector composed by m rows in which some of the original cells are summed up in a unique entry. When dealing with a square matrix of dimension n, an aggregate square matrix of dimension m can be obtained by pre-multiplying the original matrix by $\mathbf{S}$ ($m \times n$) and post-multiplying it by $\mathbf{S}'$ ($n \times m$).

The aggregation in input–output models is related to two main dimensions: the resolution of sector/commodity disaggregation of input–output matrices and related extensions and the level of spatial/geographical aggregation (Miller and Blair, 2009, p. 160).

The issues of sector/commodity aggregation in input–output models and quantification of its bias have been investigated for a long time, the first relevant article explicitly dealing with this issue dating back to 1952 (Hatanaka, 1952). The main concern at that time was related to computational constraints when dealing with big matrices. Aggregation was one way of easing the computation of the Leontief inverse. However, due to tremendous improvements in computational power, the issue of aggregation is currently related to constraints on the availability of or concerns over the quality of disaggregated data. The measurement and decomposition of the bias have been investigated by Morimoto (1970).[9] Morimoto (1970) defines the aggregation bias as 'the difference between the outputs which are derived from the aggregated system and those which are derived by aggregating the results of the original system' (p. 120).[10] The main contribution by Morimoto (1970) is related to four theorems which identify the cases in which the aggregation bias does not arise.[11] To summarize, the aggregation bias in static input–output models disappears if, alternatively: (a) the sectors/commodities which are aggregated are characterized by the same interindustry structure; (b) the vector of final demand remains unchanged for all aggregated sectors/commodities whereas it changes for all or some of the non-aggregated sectors/commodities.

However, when dealing with extensions (e.g. environmental extensions) either these conditions should be used together or the additional condition of 'common emissions coefficient among aggregated sectors/commodities' should be satisfied. A more complete discussion on the theoretical assessment of the aggregation bias in EE-IOA can be found in two recent articles by Su *et al.* (2010) and Lenzen (2011); as previously indicated Su *et al.* (2010) focus on a description of the aggregation bias and its generalization and they perform sensitivity analysis in order to identify a minimum level of disaggregation (around 40 sectors) to assure reliable estimates. Lenzen (2011) demonstrates that it is generally desirable to have approximations of disaggregated input–output relations when environmental information is available at a very disaggregated level instead of aggregating environmental information to the level of original actual input–output data.

In our case, the aggregation bias is likely to arise because, when assessing the consumer responsibility, we consider the vector of *domestic* final demand (thus excluding the vector of export) instead of *total* final demand. This is equivalent to estimating the effect of a particular impulse (different from the actual vector of final demand) with the risk of obtaining biased results.

The main purpose of the current analysis is to aggregate our original Italian and Spanish data according to relevant aggregations used in other studies and to compare our benchmark estimates (i.e. the estimates arising from the most disaggregated model) with the estimates arising from less detailed aggregations. More specifically, our benchmark consists of a disaggregation of 50 commodities.[12] This benchmark will be compared with the sub-section NACE Rev. 1.1 level (accounting for 30 sectors) and with an aggregation of 16 sectors roughly corresponding to previous studies based on OECD/IEA data sources such as Ahmad and Wyckoff (2003) and Nakano *et al.* (2009).[13] Table 8.2 summarizes the sectoral detail of each aggregation we tested.

Even if several studies acknowledge that their results depend on the choice of the level of aggregation, to our knowledge, just two of them explicitly performed a sensitivity test for aggregation bias. Wyckoff and Roop (1994) found that aggregating their analysis[14] to six sectors (using a disaggregation of 33 sectors as a benchmark) downward biases the carbon embodied in manufacturing imports by about 30 per cent. Su *et al.* (2010) perform a similar sensitivity analysis on a single

Table 8.2 Sector aggregation

Aggregation level	*Detail*
50-sector aggregation	2-digit NACE Rev. 1.1 except 50–52, 65–67 and 70–74
30-sector aggregation	Sub-sections NACE Rev. 1.1 (2-digit capital letters): A (01–02), B (05) CA (10–12), CB (13–14), DA (15–16), DB (17–18), DC (19), DD (20), DE (21–22), DF (23), DG (24), DH (25), DI (26), DJ (27–28), DK (29), DL (30–33), DM (34–35), DN (36–37), E (40–41), F (45), G (50–52), H (55), I (60–64), J (65–67), K (70–74), L (75), M (80), N (85), O (90–93), P (95)
16 sector-aggregation	Agriculture, hunting, forestry and fishing (01–05); Mining and quarrying and petroleum refining (10–14, 23); Food products, beverages and tobacco (15–16); Textiles, apparel and leather (17–19); Wood and wood products (20); Pulp, paper, printing and publishing (21–22); Chemicals (24); Other non-metallic mineral products (26); Iron and steel (271, 2731) and Non-ferrous metals (272, 2732); Fabricated metal products, machinery and equipment (28–32); Motor vehicles, trains, ships, planes (34–35); Plastics, other manufacturing and recycling (25, 33, 36–37); Electricity, gas (40); Construction (45); Transport and storage (60–62); All other services (41, 50–93 excl. 60–62)

Source: Ahmad and Wyckoff (2003).

country environmentally extended input–output model for China. Compared to their benchmark results obtained with a disaggregation of 122 sectors,[15] the bias in the estimation of carbon emissions embodied in Chinese exports arising from aggregation is positive and around 12 per cent when using a 10-sector aggregation whereas it almost vanishes when using a 42-sector aggregation.

Data sources

The current analysis relies on input–output tables for Italy and Spain for the years 1995, 2000 and 2005 with a disaggregation of 60 sectors/commodities and on NAMEA sector-level air emissions data with a disaggregation of 50 sectors for the same years and countries. To match the environmental extensions with the input–output table, we reduced the overall level of disaggregation to 50 sectors. In this section, we discuss the features and the limitations of our base data in detail.

Input–output tables

A monetary input–output table is a square matrix in which the entries in each column represent the composition (in monetary terms) of the intermediate inputs of a sector whereas the entries in each row represent how the output of a sector is used by the rest of the economic system, either as intermediate inputs or as final consumption.

The Council Regulation (EC) No. 2223/96 of 25 June 1996 on the European system of national and regional accounts in the Community (the so-called ESA 1995) requires each member country to compile and submit supply-and-use tables annually and symmetric (domestic and import) input–output tables every five years to Eurostat. The regulation is very precise as regards the methodology to be used to collect the data and the structure of the published data but allows some flexibility as regards the choice between 'commodity-by-commodity' and 'industry-by-industry' input–output tables. On the one hand, commodity-by-commodity input–output tables better describe the actual technology in terms of intermediate commodities to produce a specific product whereas industry-by-industry input–output tables describe relationships among sectors regardless of the actual flows of commodities. On the other hand, most of the extensions (e.g. environmental extensions) refer to industries and not to commodities, making the 'industry-by-industry' approach more attractive (Eurostat, 2008; Miller and Blair, 2009). Out of the 31 countries which submit their input–output tables to Eurostat (EU27 plus Croatia, Macedonia, Turkey and Norway), 'industry-by-industry' tables are only supplied by eight countries (Denmark, Italy, Hungary, Netherlands, Finland, UK, Turkey and Norway).

In our analysis, we use 'commodity-by-commodity' input–output tables in order to make the comparison between Italy and Spain possible. The procedure we use to assign 'industry' emissions to 'commodity' output is based on the hypothesis that direct emissions related to each commodity within a single industry are proportional to the share of the output of each commodity within the industry (Miller and Blair, 2009). Information on the commodity composition of industry output can be found in the make (supply) matrix.

Starting with the make matrix (**V**) and the vector of total output by industry (**x**), we compute a matrix which describes the commodity composition of industry output ($\mathbf{C}=\mathbf{V}'\langle\mathbf{x}\rangle^{-1}$). Each row of the matrix sums to 1 and indicates the relative weight of the different commodities in the total output of the industry (Roca and Serrano, 2007; Miller and Blair, 2009).[16] To obtain the measure of direct emissions generated by the production of a specific commodity (by all of the industries producing that commodity), indicated with $\mathbf{e}_{pp}$, we multiply the transpose of **C** by the vector of direct emissions by industry ($\mathbf{e}_{ii}$):

$$\mathbf{e}_{pp} = \mathbf{C}'\mathbf{e}_{ii} \tag{3}$$

In Appendix A we compare our results obtained using the commodity-by-commodity approach for Italy with the results we obtain using the industry-by-industry approach.[17] The estimates for total emissions induced by domestic final demand differ by less than 1 per cent in all cases except for CO in 2000 and 2005, thus confirming the validity of the 'commodity-by-commodity' approach.

NAMEA data

The NAMEA approach to identify environmental pressures across production sectors was developed in the late 1980s and 1990s at the Central Bureau of

Statistics of the Netherlands (CBS) under the supervision of Steven Keuning (De Boo *et al*., 1991). As widely presented above,[18] NAMEA data are constituted by a matrix form statistical source where economic (output, value added, final consumption expenditures and full-time job equivalent) and environmental (emissions) indicators can be observed at sector level. In NAMEA, environmentally relevant information is compiled consistently with the way economic activities are represented in national accounts. This framework divides the economy into production sectors and household consumption categories and shows how each industry branch or the household categories contribute to a set of environmental pressures. This allows quite robust analyses on dynamics, correlation, even causation regarding performance and resource productivity indicators. De Haan and Keuning (1996) and Stauvermann (2007), among others, are examples of seminal papers containing long and comprehensive bibliographies of all past works. Furthermore, de Haan (2004) developed and propagated the NAMEA approach in detail and has applied NAMEA for international comparisons.

In the NAMEA tables, environmental pressures (for Italian NAMEA air emissions and virgin material withdrawal) and economic data are assigned to the economic branches of productive resident units or to the household consumption categories (heating, transport and others) directly responsible for environmental and economic phenomena. We focus here on production macro sectors – agriculture and fishing, industry and services – obtained by aggregating the available economic branches at national level to capture the main potential differences in environmental pressure.

Both the Italian, which dates back to 1990 (first published data in 2000), and the Spanish NAMEA include several air pollutants: carbon dioxide (CO_2), nitrogen oxides (NO_x), methane (CH_4), sulphur oxides (SO_x), nitrous oxide (N_2O), ammonia (NH_3), non-methane volatile organic compounds (NMVOC) and carbon monoxide (CO) among others. In the current chapter, we report results for emissions of five different substances (CO_2, NO_x, SO_x, NMVOC, CO)[19] for which NAMEA with the same aggregation of sectors is available both for Italy and Spain.[20]

Results and discussion

Overview: consumption vs. production perspective in the benchmark case

Before facing the issue of aggregation and its related bias, in this section we briefly discuss the results for Italy and Spain of our benchmark (50 sectors) estimates for the years 1995 and 2005. The 50-sector aggregation level has been obviously considered as the benchmark; as stated by Su *et al*. (2010), in empirical studies it is logical to take the view that the finer the level of sector disaggregation, the more refined the decomposition results obtained.

Figures 8.1–8.2 and 8.3–8.4 report the contribution of three macro-sectors[21] to emissions induced by domestic final demand and domestic direct emissions for Italy and Spain respectively.

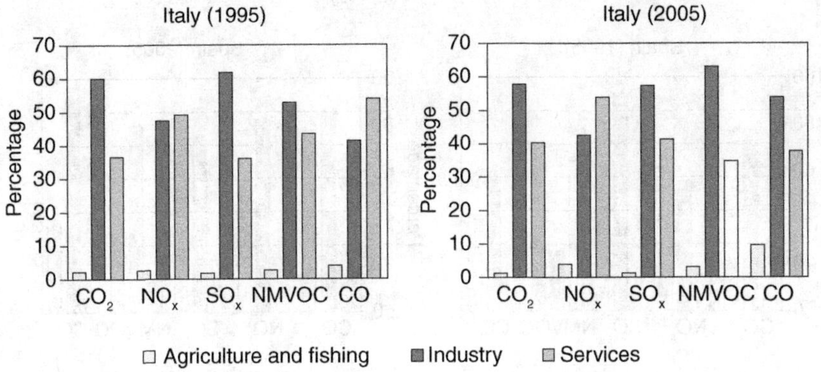

Figure 8.1 Emissions induced by domestic final demand by sector (Italy).

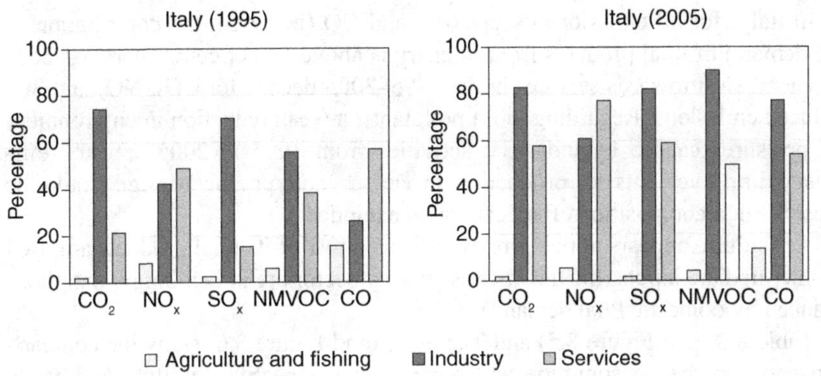

Figure 8.2 Direct emissions by sector (Italy).

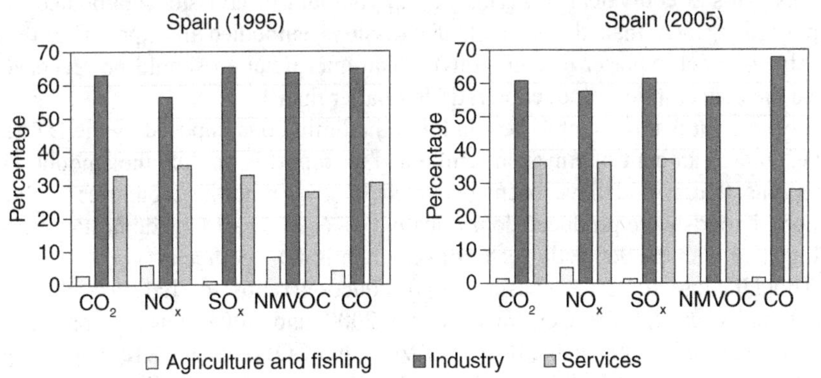

Figure 8.3 Emissions induced by domestic final demand by sector (Spain).

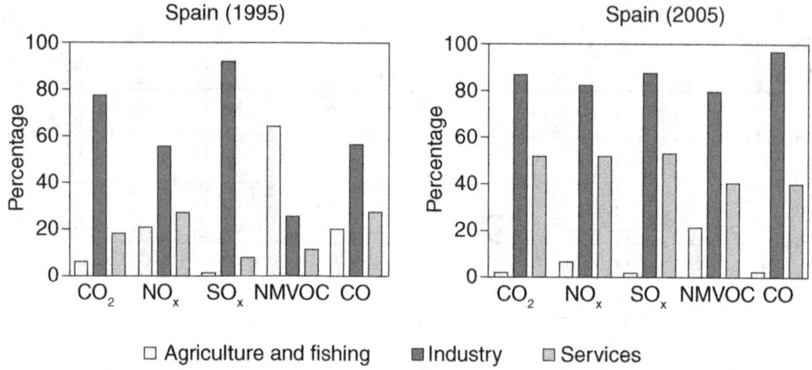

Figure 8.4 Direct emissions by sector (Spain).

In Italy, for all emissions except NO_x and CO (in 1995), the contribution of the demand of final products from industry is above 50 per cent. There has been a general shift towards services in the 1995–2005 decade for CO_2, NO_x and SO_x induced emissions. Regarding those pollutants, a weak reduction in environmental pressures caused by industrial activities from 1995 to 2005 appears; efficiency improvements in production processes and product design could be present but a composition effect cannot be excluded.

Agriculture appears almost irrelevant since most of its final products are used as intermediate inputs (the direct emissions by sector are in fact bigger that those induced by domestic final demand).

Table 8.3 (and Figure 8.5) and Table 8.4 (and Figure 8.6) show the comparison between the consumption and production perspective for Italy and Spain respectively. A consumption/production ratio greater than 1 indicates that the emissions arising from the production needed to satisfy the domestic final demand are greater than the emissions directly generated by domestic production sectors. This is equivalent to saying that the amount of emissions embodied in imports is greater than the amount of emissions embodied in export (i.e. the country is a net exporter of emissions).[22] The interpretation should be reversed when the consumption/production ratio is smaller than 1.

However, it is important to bear in mind the limitations imposed by the DTA: the estimated amount of emissions embodied in import is, in fact, the amount of emissions that would have been generated by the importing countries if the imported goods were produced domestically (i.e. according to the domestic technology and with the domestic vector of sector emission coefficients).

Though close to 1, the consumption/production ratios for Italy are always below unity except for CO emissions in 2000 and 2005. Furthermore, the average pattern is either stable (CO_2, NMVOC and CO) or even decreasing (NO_x and SO_x). This result, in line with previous analyses such as Moll *et al.* (2007) but still quite surprising for an OECD country, may have two main explanations.

Table 8.3 Emissions for production and consumption perspective (Italy, 50 sectors; in tons, CO_2 in 1,000 tons)

	Production perspective			*Consumption perspective*		
	1995	*2000*	*2005*	*1995*	*2000*	*2005*
CO_2	360,071	368,511	389,961	348,183	355,362	376,104
NO_x	1,569,712	1,233,273	1,139,097	1,507,256	1,132,557	1,035,779
SO_x	1,375,635	840,127	457,795	1,374,334	774,669	398,884
NMVOC	1,064,689	713,566	584,124	1,002,686	670,275	557,370
CO	3,034,181	1,539,949	1,212,926	2,965,820	1,559,251	1,232,689

First, Italy maintained industrial specialization in the manufacturing sector, especially in more traditional (and relatively energy intensive) industries, during the considered period. Second, it may be that, within each two-digit industry, there has been a shift from polluting sub-industries (whose products, formerly produced domestically, have been substituted by import) to cleaner sub-industries. This possible shift may lead to a reduction in direct sector emissions in presence of unchanged aggregate monetary domestic output (though with a different sub-industry composition not visible in aggregate monetary data), thus artificially improving the environmental efficiency of the aggregate sector. This hidden structural change worsens the DTA prediction because it affects the sub-industry composition and the real average environmental efficiency of imports. This possible explanation further highlights the importance of using disaggregate data.

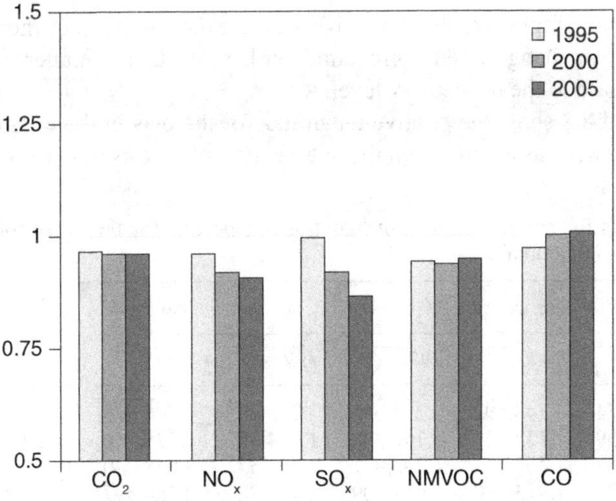

Figure 8.5 Consumption/production perspective (Italy, 50 sectors).

The comparison between the patterns of different emissions suggests other somewhat unexpected and interesting results. Local negative externalities generated by NO_x and SO_x (and not by CO_2) emissions, coupled with relatively strict environmental policies controlling these emissions during the considered period,[23] are expected to increase the incentive to move the production of commodities intensive in these emissions abroad (to pollution havens). This should result in an increase of emissions embodied in imports and an increase in the consumption/production ratio. However, we find the opposite which suggests that Italy, due to low stringency of environmental regulation and to lack of enforcement, is behaving as a pollution haven within the EU (Marin and Mazzanti, in press).

Spain is characterized by the opposite situation and pattern. For all emissions/years the consumption/production ratio is greater (often far greater) than 1 and the ratio tends to increase in time, reaching the maximum for SO_x in 2000 with 1.395. This means that SO_x emissions induced by domestic final demand are 39.5 per cent greater than SO_x emissions directly generated by Spanish industries. These results are in line with the findings of Arto *et al.* (2010) and Serrano and Dietzenbacher (2010).

Spain had a very dynamic economy during the 1990s and the early 2000s, with growth mainly driven by the construction and tertiary sectors whereas the share of manufacturing in employment, output and value added has declined steadily.[24] This process, coupled with an increased volume of final demand of manufacturing goods (Roca and Serrano, 2007), gave rise to a rapid increase in foreign emissions to produce these goods, thus worsening the balance of emissions embodied in import.

Aggregation bias

In the following paragraphs, we discuss to what extent the estimates of the consumption perspective change when aggregating our base data, in a smaller subset of sector, with regard to the benchmark level.

Figures 8.7 and 8.8 show the relative magnitude of the bias in the consumption perspective emissions arising from the aggregation of sectors into 30 NACE

Table 8.4 Emissions for production and consumption perspective (Spain, 50 sectors; in tons, CO_2 in 1,000 tons)

	Production perspective			*Consumption perspective*		
	1995	*2000*	*2005*	*1995*	*2000*	*2005*
CO_2	208,054	248,692	294,655	220,225	306,978	382,698
NO_x	1,028,209	1,155,724	1,257,268	1,074,762	1,328,240	1,560,148
SO_x	1,752,362	1,453,493	1,290,977	1,891,531	2,028,020	1,750,648
NMVOC	1,865,274	1,913,460	1,987,809	2,181,989	2,380,397	2,453,815
CO	908,522	932,967	904,531	993,401	1,158,443	1,243,147

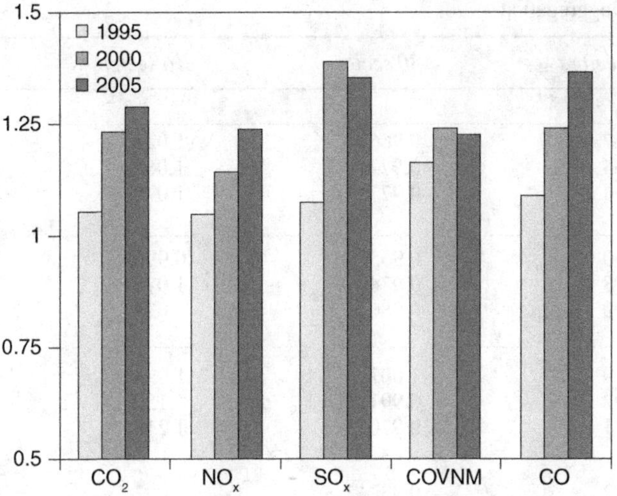

Figure 8.6 Consumption/production perspective (Spain, 50 sectors).

Rev 1.1 (see also Appendix B and Table 8.A.1) sub-sections and in 16 sectors according to the IEA/OECD studies[25] in the Italian case.

First note that, with few exceptions (CO_2 in 1995 and CO in 1995 and 2000 for the 30-sector aggregation), aggregation tends to overestimate the consumption perspective and this effect is even more evident in the 16-sector aggregation. Moreover, the bias tends to increase in time. The bias tends to be greater for the 16-sector aggregation as opposed to the 30-sector aggregation.[26]

With regard to the 16-sector aggregation, the magnitude of the bias is particularly evident for SO_x (with a maximum bias of almost 40 per cent in 2005) and it is also relevant for NMVOC, CO_2 and NO_x.

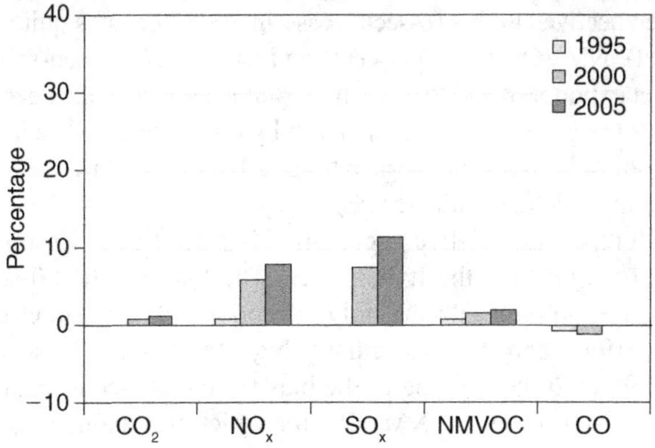

Figure 8.7 Aggregation bias %: 30 vs. 50 sectors (Italy).

Table 8.5 Consumption/production perspective emissions for Italy according to different levels of aggregation

Year	*50 sectors*	*30 sectors*	*16 sectors*
CO_2			
1995	0.967	0.966	1.021
2000	0.964	0.972	1.067
2005	0.964	0.977	1.077
NO_x			
1995	0.960	0.965	0.990
2000	0.918	0.974	1.016
2005	0.909	0.980	1.027
SO_x			
1995	0.999	1.001	1.093
2000	0.922	0.991	1.150
2005	0.871	0.970	1.216
NMVOC			
1995	0.942	0.952	1.003
2000	0.939	0.956	1.035
2005	0.954	0.973	1.079
CO			
1995	0.977	0.970	1.006
2000	1.013	1.004	1.072
2005	1.016	1.016	1.091

The detailed estimates of the consumption/production perspective ratio for the different levels of aggregation (Table 8.5) show to what extent the aggregation bias is likely to affect our main synthetic indicator, the consumption/production perspective ratio. In all cases (again except CO), moving from the benchmark result (50 sectors) to the result for 16 sectors (to be compared with the set of IEA/OECD multi-regional analyses) artificially makes Italy a net exporter of emissions even within the framework of a pure DTA. Moreover, the relative gap between consumption and production perspectives in the 16-sector case in 2005 becomes quite high for SO_x (+21.6 per cent), NMVOC (+7.9 per cent) and CO_2 (+7.7 per cent),[27] suggesting that Italy is a net exporter of emissions. This result is just an artefact because the choice of the sectors to aggregate, although driven by constraints in data availability in IEA/OECD studies, is arbitrary. We also tried different apparently reasonable aggregation and obtained quite volatile results.

Figures 8.9 and 8.10 report the relative aggregation bias for Spain. Results for Spain are less straightforward than the Italian ones. The bias for the 30-sector aggregation is generally negative (with the only exceptions of very small positive biases for NO_x in 1995) and it is particularly high for NMVOC. No clear trend is found from 1995 to 2005. Moving to the bias for the 16-sector aggregation, it is generally positive (except for NMVOC for which it remains negative, though less important relative to the 30-sector aggregation). Moreover, it tends to decrease in time for CO_2, NO_x and SO_x and to increase for CO.

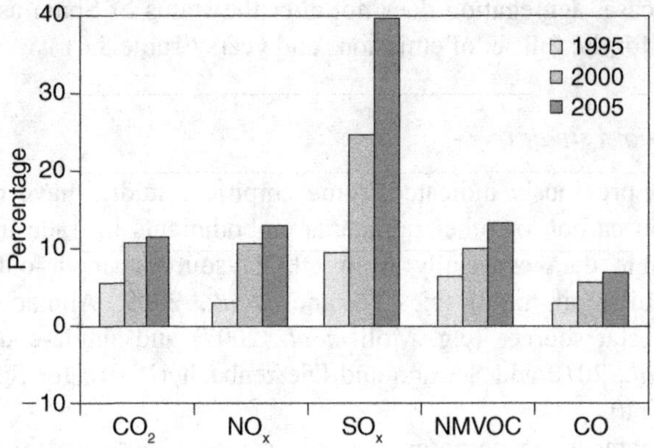

Figure 8.8 Aggregation bias %: 16 vs. 50 sectors (Italy).

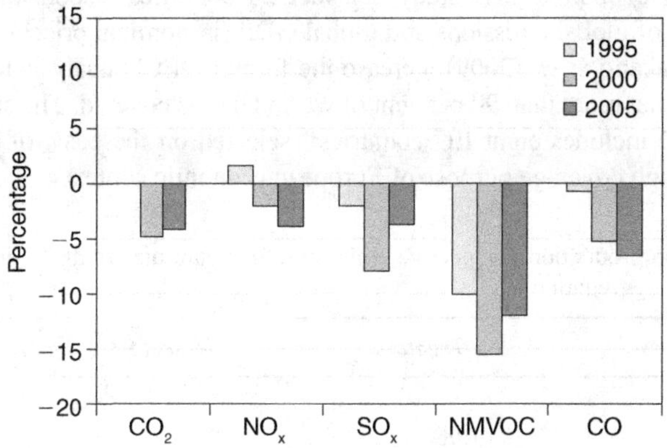

Figure 8.9 Aggregation bias %: 30 vs. 50 sectors (Spain).

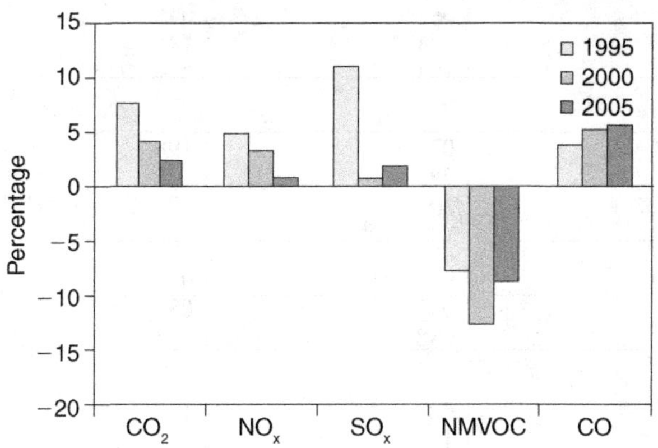

Figure 8.10 Aggregation bias %: 16 vs. 50 sectors (Spain).

Unlike the Italian case, aggregation does not alter the status of Spain as net exporter of emissions for the full set of emissions and years (Table 8.6).

Comparison with previous studies

In the last decade, as previously indicated, some empirical studies have been conducted focusing on carbon or other pollutants embodiments in trade using international-comparable data especially from OECD sources (input–output, CO_2 emissions and bilateral trade) (e.g. Nakano *et al.*, 2009; Ahmad and Wyckoff, 2003), Eurostat sources (e.g. Moll *et al.*, 2007) and single-country sources (e.g. Arto *et al.*, 2010 and Serrano and Dietzenbacher, 2010 for Spain; Su *et al.*, 2010 for China).

Among them, for comparison purposes, we only consider those that include Italy or Spain or both. However, it is important to underline that some of the considered studies were defined with a global and cross-country analysis focus. Ahmad and Wyckoff (2003) in their OECD study consider 24 countries responsible in 1995 for 80 per cent of global emissions and global GDP (in nominal prices); following this study, Nakano *et al.* (2009) increase the former OECD analysis to 41 countries/regions so that more than 90 per cent of world GDP is covered. The study of Moll *et al.* (2007) includes eight EU countries[28] selected on the basis of data availability and the high coverage purpose of European economic contexts.

Table 8.6 Consumption/production perspective emissions for Spain according to different levels of aggregation

Year	*50 sectors*	*30 sectors*	*16 sectors*
CO_2			
1995	1.059	1.060	1.142
2000	1.234	1.176	1.288
2005	1.299	1.242	1.331
NO_x			
1995	1.045	1.062	1.096
2000	1.149	1.123	1.186
2005	1.241	1.193	1.249
SO_x			
1995	1.079	1.079	1.198
2000	1.395	1.285	1.405
2005	1.356	1.301	1.383
NMVOC			
1995	1.170	1.049	1.079
2000	1.244	1.047	1.084
2005	1.234	1.088	1.125
CO			
1995	1.093	1.083	1.137
2000	1.242	1.179	1.306
2005	1.374	1.283	1.453

A comparison of our CO_2 results with the empirical evidence for the same pollutant found in the recent EE-IOA studies suggests that as far as the Italian case is concerned (Table 8.7), some literature references could be affected by aggregation bias due to a small number of considered sectors. This results in a strong and significant difference among empirical findings with respect to both the consumption and production perspective emissions and the corresponding ratio. In Nakano *et al.* (2009) and Ahmad and Wyckoff (2003),[29] the C/P ratios reported for the Italian case, in 1995 and 2000, are larger than ours and always higher than 1. The Moll *et al.* (2007) figure is the closest to our 2000 figure for the C/P ratio (0.96); they use a 38-sector aggregation level and if we consider the sensitive results found by Su *et al.* (2010) and discussed in the second paragraph (levels around 40 sectors appear to be sufficient to capture the overall share of emissions embodied in a country's export), it may be considered more reliable than other authors' findings. From a policy point of view, a C/P ratio that ranges from 1.24–1.30 to 0.96–0.97 suggests that, while large studies that involve several countries have to be encouraged because they permit macro area analysis, if they require a low level of sectoral detail to ensure countries' homogeneity and comparability, their empirical results require caution when they are interpreted.

Table 8.8 shows a similar comparison for Spain. With regard to this country, the empirical findings reported in the different studies are more homogeneous than the Italian case both for the absolute values of production and consumption perspective CO_2 emissions and the corresponding ratio. This could be interpreted, at least partially, as a confirmation of the reliability of our 50-sector estimates. However, in the light of the Italian results, we could conclude that after a certain degree of aggregation, there is a concrete risk of having biased and volatile results which depend on the specificities of the economic structure of the country and the type of emission considered.

Table 8.7 CO_2 emissions for production and consumption perspective in Italy in different studies (Mton CO_2)

Source[1]	*Italy*					
	Production perspective		*Consumption perspective*		*C/P*	
	1995	*2000*	*1995*	*2000*	*1995*	*2000*
Nakano *et al.* (2009), MRIO, 17	413	427	511	554	1.24	1.30
Ahamd and Wyckoff (2003) MRIO, 17	398[2]		445[2]		1.12	
Moll *et al.* (2006) DTA, 38		358		362		1.01
Own elaboration DTA, 50	360	369	348	355	0.97	0.96

Notes
1 Author(s), year, DTA or MRIO, aggregation level (#sectors);
2 1992.

Table 8.8 CO_2 emissions for production and consumption perspective in Spain in different studies (Mton CO_2)

Source[1]	*Spain*					
	Production perspective		*Consumption perspective*		*C/P*	
	1995	*2000*	*1995*	*2000*	*1995*	*2000*
Nakano *et al.* (2009), MRIO, 17	236	280	275	330	1.17	1.18
Ahamd and Wyckoff (2003), MRIO, 17	235		252		1.07	
Serrano and Dietzenbacher (2010), DTA, 46	204	239	222	279	1.09	1.17
Arto *et al.* (2010), DTA, 46 sect		364[2]		429[3] 453[4]		1.18 1.24
Moll *et al.* (2006), DTA, 46	209		228		1.09	
Own elaboration DTA, 50	208	249	220	307	1.06	1.23

Notes
1 Author(s), year, DTA or MRIO, aggregation level (#sectors);
2 $MtCO_2e$;
3 $MtCO_2e$ with Monetary DTA;
4 $MtCO_2e$ with Physical DTA.

Conclusions

The integration of the National Accounting Matrix including Environmental Accounts (NAMEA) and input output (I–O) tables (often referred to as Environmentally extended Input–Output Analysis – EE-IOA – based on NAMEA data) represents a new way to analyse the determinants of the income–environment relationships in international settings. Moreover, EE-IOA provides analyses of the emissions embodied in domestic consumption and domestic production by considering the structure of intermediate inputs and environmental efficiency in each production sector.

A comparison of a production and consumption perspective may have relevant policy implications. A consumption and production emission ratio greater than 1 denotes a country that is a net exporter of emissions in the sense that it requires an amount of emissions embodied in imports, and thus produced abroad, that is greater than the amount of emissions embodied in export. Usually, the environmental policy points mainly to production activities as responsible actors of impacts to be targeted by legislation and regulation. Looking at the final consumption demand for vertically integrated domestic and international environmental impacts can push policy attention towards the possible role of consumers as actors to be targeted with particular environmental policies, together with the international responsibility for environmental externalities of pollutants' emissions produced abroad but domestically demanded.

However, similar comparisons require particular assumptions, such as the technology associated with the imported goods, and could be affected by some biases. In this chapter we have analysed and discussed the aggregation bias due to different levels of production sector aggregation for Italy and Spain in 1995, 2000 and 2005. Our empirical findings, for the Italian and the Spanish cases, show that different sectoral aggregation significantly biases the amount of emissions both for the consumption and the production perspective. At the level where we consider only 16 production sectors, the results obtained in both the consumption and production perspective are quite different from those for higher levels of sector disaggregation (e.g. 50 which is our benchmark) both for the amounts of calculated emissions and for the corresponding C/P ratios. With regard to Italy, the 16-sector aggregation level in 2005 shows an emissions amount for CO_2, NO_x and NMVOC which is more than 10 per cent higher than those calculated with the 50-sector aggregation level. Moreover, considering SO_x, the gap between 16- and 50-sector aggregation reaches almost 40 per cent. With regard to Spain, between 16- and 50-sector aggregation levels in 2005, there are differences of below +5 per cent for CO_2, NO_x and SO_x, and almost 5 per cent for CO. NMVOC shows the biggest gap for the Spanish case with an under-estimation of almost –8 per cent compared with the benchmark aggregation level due to the use of a 16-sector aggregation level.

Our results suggest that special attention must be paid when interpreting the EE-IOA of country estimated amounts of embodied emissions, both in domestic final demand and those directly associated with the production sectors when the sectoral aggregation level has a low definition as considered in some recent similar studies.

Appendix A

The methodology we used to employ in a consistent way commodity-by-commodity input–output tables as a proxy of industry-by-industry tables is explained above. While the main analysis relies on results obtained using commodity-by-commodity input–output tables, in this appendix we report the differences between the industry-by-industry approach and the commodity-by-commodity approach as regards the estimation of the emissions induced by domestic demand. This comparison is only possible for Italy because Spain does not publish industry-by-industry input–output tables. The main results are summarized in Table 8.A.1.

With the only exception of CO emissions, the absolute value of the gap for aggregate consumption perspective emissions is always below 1 per cent. On average, the commodity-by-commodity approach tends to under-estimate the emissions induced by the final demand of agriculture/fishing goods and industrial goods whereas it over-estimates the emissions induced by the final demand of services. Finally, we do not observe relevant changes in the magnitude of the gaps over time.

Table 8.A.1 Commodity-by-commodity (cc) versus industry-by-industry (ii) approach for Italy (1-ii/cc) (%)

1995	CO_2	NO_x	SO_x	*NMVOC*	*CO*
Agriculture and fishing	−4.74	−4.64	−4.34	−4.34	−5.41
Industry	−2.79	−0.83	−2.33	−2.08	−1.09
Services	2.60	−0.70	1.94	1.66	−0.14
Total	−0.84	−0.88	−0.83	−0.48	−0.74
2000	CO_2	NO_x	SO_x	*NMVOC*	*CO*
Agriculture and fishing	−4.61	−4.37	−4.54	−4.63	−5.25
Industry	−4.17	−2.42	−3.37	−4.37	−5.93
Services	5.95	3.24	5.87	7.62	4.95
Total	−0.55	0.34	−0.02	−0.41	−1.61
2005	CO_2	NO_x	SO_x	*NMVOC*	*CO*
Agriculture and fishing	−4.69	−4.57	−4.47	−5.36	−6.17
Industry	−3.91	−2.52	−3.36	−4.27	−6.48
Services	4.57	2.48	4.17	7.62	6.71
Total	−0.63	0.02	−0.37	−0.51	−1.88

Appendix B

Table 8.B.1 NACE Rev. 1.1; 2-digit

01 Agriculture, hunting and related service activities
02 Forestry, logging and related service activities
05 Fishing, fish farming and related service activities
10 Mining of coal and lignite; extraction of peat
11 Extraction of crude petroleum and natural gas; service activities incidental to oil and gas extraction, excluding surveying
12 Mining of uranium and thorium ores
13 Mining of metal ores
14 Other mining and quarrying
15 Manufacture of food products and beverages
16 Manufacture of tobacco products
17 Manufacture of textiles
18 Manufacture of wearing apparel; dressing and dyeing of fur
19 Tanning and dressing of leather; manufacture of luggage, handbags, saddlery, harness and footwear
20 Manufacture of wood and of products of wood and cork, except furniture; manufacture of articles of straw and plaiting materials
21 Manufacture of pulp, paper and paper products
22 Publishing, printing and reproduction of recorded media
23 Manufacture of coke, refined petroleum products and nuclear fuel
24 Manufacture of chemicals and chemical products
25 Manufacture of rubber and plastic products
26 Manufacture of other non-metallic mineral products
27 Manufacture of basic metals
28 Manufacture of fabricated metal products, except machinery and equipment
29 Manufacture of machinery and equipment n.e.c.
30 Manufacture of office machinery and computers

31 Manufacture of electrical machinery and apparatus n.e.c.
32 Manufacture of radio, television and communication equipment and apparatus
33 Manufacture of medical, precision and optical instruments, watches and clocks
34 Manufacture of motor vehicles, trailers and semi-trailers
35 Manufacture of other transport equipment
36 Manufacture of furniture; manufacturing n.e.c.
37 Recycling
40 Electricity, gas, steam and hot water supply
41 Collection, purification and distribution of water
45 Construction
50 Sale, maintenance and repair of motor vehicles and motorcycles; retail sale of automotive fuel
51 Wholesale trade and commission trade, except of motor vehicles and motorcycles
52 Retail trade, except of motor vehicles and motorcycles; repair of personal and household goods
55 Hotels and restaurants
60 Land transport; transport via pipelines
61 Water transport
62 Air transport
63 Supporting and auxiliary transport activities; activities of travel agencies
64 Post and telecommunications
65 Financial intermediation, except insurance and pension funding
66 Insurance and pension funding, except compulsory social security
67 Activities auxiliary to financial intermediation
70 Real estate activities
71 Renting of machinery and equipment without operator and of personal and household goods
72 Computer and related activities
73 Research and development
74 Other business activities
75 Public administration and defence; compulsory social security
80 Education
85 Health and social work
90 Sewage and refuse disposal, sanitation and similar activities
91 Activities of membership organizations n.e.c.
92 Recreational, cultural and sporting activities
93 Other service activities

Acknowledgements

The authors thank Roberto Zoboli for its very useful comments and suggestions. Roberto presented a very preliminary version of this work to the conference 'The Structure of Economic Systems Through Input–Output Applications' (Accademia dei Lincei, Rome, Italy) in October 2010. The authors also want to acknowledge the Environmental Accounting Unit of ISTAT (Italy) for its very valuable work in regularly providing the NAMEA series. The usual disclaimer applies.

Notes

1 Some main streams of research can be outlined: I–O models accounting for trade and embodied emissions (through energy accounts); global multi-region input–output

(MRIO) model; extension for eco-footprint analysis; comparing physical trade balance (PTB) and pollution trade balance (UTB) associated with fossil use; analysing pollution terms of trade, pollution haven tests; analysing I–O tables linked with satellite accounts. For brevity, we refer the reader to the mounting, extensive literature that is also touched on by many contributions in this volume.

2 The physical DTA refers to the use of imports in physical quantities and using, for imports, the same physical environmental coefficient (emissions per kg of import) as domestic physical environmental coefficients (emissions per kg of domestic output). This assumes that, although of different quality (value per physical unit), the emissions content of goods is closely correlated to its weight and less correlated to its value.

3 Automobile driving and related maintenance activities are by far the largest contributing products to total environmental impacts by consumption in the EU25. However, by summing several animal-based foods (meat, meat products, poultry, dairy products), animal food products would become dominant. At the aggregate level of 12 consumption domains, food already emerges as the largest contributor to environmental problems.

4 The *consumption-based* emissions are computed using production-based emissions minus the emissions embodied in exports plus those embodied in imports.

5 For an exhaustive review on the accounting definitions related to EE-IOA, the reader is referred to Serrano and Dietzenbacher (2010) and Moll *et al.* (2007).

6 In this section we refer to the aggregation of basic data as opposed to the aggregation of results. The aggregation of results of any empirical analysis in economics is a necessary step when giving an overall picture of the phenomenon under analysis.

7 Due to privacy protection, ISTAT, the Italian National Institute of Statistics, is not allowed to publish data for aggregates with less than three units and it is forced to further aggregate these branches.

8 The aggregation to the minimum common standard is the most widely used approach (Ahmad and Wyckoff, 2003; Nakano *et al.*, 2009). However, a noticeable exception is represented by Huppes *et al.* (2005) who exploit the very detailed US input–output table and adapt it to the EU economic structure, thus using more disaggregated data relative to publicly available EU input–output tables. Although very interesting, this approach is affected by problems related to differences between US and EU classification structures within each macro-industry.

9 The theoretical results obtained by Morimoto (1970) do not depend on the reason that induces aggregation.

10 In our case, the aggregation bias is defined in terms of *emissions induced by domestic demand*. However, for the sake of brevity, we will discuss overall aggregation bias only. Results for the assessment of the aggregation bias at sector/commodity level are available upon request.

11 An important point, which often remains implicit, is that the aggregation bias only arises when the vector of final demand is modified relative to the original vector of final demand.

12 This level of disaggregation corresponds roughly to the two-digit NACE Rev. 1.1 classification (see Table 8.B.1 for a description of each sector).

13 OECD/IEA estimates use a disaggregation of 17 sectors. However, both OECD input–output tables and IEA CO_2 emissions from fuel combustion go beyond the two-digit NACE Rev. 1.1 as regards sector 27. This sector is split into 'Iron and steel' (271+2731) and 'Non-ferrous metals' (272+2732). On the contrary, Italian and Spanish input–output tables and NAMEA do not allow this separation.

14 They employ an environmentally extended MRIO model for six OECD countries (US, Canada, France, Germany, Japan and UK) to estimate the embodiment of carbon in imports of manufacturing products.

15 Note that the benchmark results are obtained by 'disaggregating' the original vector of emissions intensities (42 sectors) in order to meet the 122-sector aggregation of the input–output tables. This operation is likely to partly affect the reliability of the estimates for the 122-sector aggregation.

16 Note that when the make matrix is diagonal (that is, when all industries produce only their primary commodity), then the **C** matrix is an identity matrix.

17 This comparison is not feasible for Spain because the Instituto Nacional de Estadistica (INE) does not produce industry-by-industry input–output tables.

18 For an overview of the methodological issues related to NAMEA, we also refer the reader to Chapters 1 and 2.

19 We also perform all the estimates for 12 additional substances available in the Italian NAMEA only (NH_3, PM_{10}, $PM_{2.5}$, As, Cd, Cr, Cu, Hg, Ni, Pb, Se and Zn). Results are available upon request.

20 The Spanish NAMEA used in this chapter is available on the Eurostat website with a 50-sector aggregation and only five pollutants. The INE divulgates a NAMEA with even more pollutants but with only 30 sectors and for this reason is not useful for our purposes.

21 Agriculture and fishing (A–B NACE Rev. 1.1), Industry (C–F NACE Rev. 1.1) and Services (G–O NACE Rev. 1.1). Results at two-digit NACE are available upon request.

22 The equivalence is explained in Serrano and Dietzenbacher (2010).

23 Among others, at EU level, the Council Directive 1980/779/EC substituted by the Council Directive 1999/30/EC of 22 April 1999 'relating to limit values for sulphur dioxide, nitrogen dioxide and oxides of nitrogen, particulate matter and lead in ambient air', the Council Directive 85/203/EEC of 7 March 1985 'on air quality standards for nitrogen dioxide', as last amended by Council Directive 85/580/EEC and the Council Directive 1999/13/EC 'on the limitation of emissions of volatile organic compounds due to the use of organic solvents in certain activities and installations'.

24 The output share of manufacturing was 32.6 per cent, 31.1 per cent and 26.7 per cent in 1995, 2000 and 2005 respectively.

25 IEA/OECD studies such as Nakano *et al.* (2009) and Ahmad and Wyckoff (2003) use a disaggregation of 17 sectors which, for sector 27 (Manufacture of basic metals), goes beyond the two-digit detail. IEA/OECD data distinguish between 'Iron and steel' (27.1 and 27.31) and 'Non-ferrous metals' (27.2 and 27.32). On the contrary, input–output tables and NAMEA published by ISTAT and INE treat sector 27 as a unique sector. This aggregation potentially introduces a bias in our results due to the high emissions intensity of sector 27 and to the heterogeneity in technologies and emissions intensity within sector 27.

26 Note that there is no perfect link between the 16-sector aggregation and the 30-sector aggregation. This fact does not allow the monotonicity of the bias with respect to the number of sectors to be interpreted as a stylized fact. In fact, monotonicity is not found for Spain.

27 The figure for the benchmark case of the 50-sector disaggregation was of –12.9 percent for SO_x, –4.6 percent for NMVOC and –3.6 percent for CO_2.

28 The selected eight economies represent more than two-thirds of EU25's GDP and more than 60 per cent of EU25's population. The geographical coverage comprises Spain, UK (1995) and Germany, Denmark, Hungary, Italy, the Netherlands, Sweden (1995 and 2000).

29 A consistent comparison of the absolute levels of CO_2 emissions between IEA/OECD studies and NAMEA-based studies is not possible. In fact, IEA records CO_2 emissions from fuel combustion only and, differently from NAMEA, the principle of recording the emissions generated by resident agents only is not applied in the collection of these data.

References

Ahmad, N. and Wyckoff, A. (2003) 'Carbon dioxide emissions embodied in international trade of goods', OECD Science, Technology and Industry Working Papers 2003/15, Paris: OECD.

Alcántara, V. and Padilla, E. (2009) 'Input–output subsystems and pollution: an application to the service sector and CO_2 emissions in Spain', *Ecological Economics*, 69: 905–914.

Andrew, R., Peters, G.P. and Lennox, J. (2009) 'Approximation and regional aggregation in multi-regional input–output analysis for national carbon footprint accounting', *Economic Systems Research*, 21: 311–335.

Arto, I., Roca, J. and Serrano, M. (2010) 'Accounting for embodied emissions in international trade: physical vs monetary trade flows', mimeo.

Basina, A. and Sfetsos, A. (2010) 'Environmental impacts of generating electricity by substituting lignite with photovoltaic technology: an analysis based on a NAMEA-table for the Greek economy', paper presented at the I–O Society Conference, Sydney, 20–25 June.

Casler, S.D. and Rose, A. (1998) 'Carbon dioxide emissions in the U.S. economy', *Environmental and Resource Economics*, 11: 349–363.

Cole, M.A. (2004) 'Trade, the pollution haven hypothesis and the environmental Kuznets curve: examining the linkages', *Ecological Economics*, 48(1): 71–81.

Copeland, B.R. and Taylor, M.S. (2004) 'Trade, growth and the environment', *Journal of Economic Literature*, 42: 7–71.

De Boo, A., Bosch, P., Gorter, C.N. and Keuning, S.J. (1991) 'An environmental module and the complete system of national accounts', Occasional Papers of the CBS, No. NA-046, Voorburg.

de Haan, M. and Keuning, S.J. (1996) 'Taking the environment into account: the NAMEA approach', *Review of Income and Wealth*, 42: 131–148.

de Haan, M. (2004) 'Accounting for goods and for bads: measuring environmental pressure in a national accounts framework', PhD thesis, Universiteit Twente.

Eurostat (2008) 'Eurostat manual of supply, use and input–output tables', Eurostat Methodologies and Working Papers, Luxembourg: Eurostat.

Frankel, J. and Rose, A. (2005) 'Is trade good or bad for the environment?' *Review of Economics and Statistics*, 87: 85–91.

Gallego, B. and Lenzen, M. (2005) 'A consistent input–output formulation of shared producer and consumer responsibility', *Economic Systems Research*, 17: 365–391.

Harris, R. (2001) 'Methods for estimating air emissions from the production of goods imported into the UK', prepared for DG Regional Policy and Eurostat, Working Paper 2/2001/B/5, Eurostat, Brussels: European Commission.

Hatanaka, M. (1952) 'Note on consolidation within a Leontief system', *Econometrica*, 20: 301–303.

Hoekstra, R. (2010) 'A complete overview of environmental input–output articles', paper presented at the I–O Society Conference, Sydney, 20–25 June.

Hoekstra, R., Smits, J.P. and van der Helm, R. (2010) 'Economic growth, structural change and environmental pressures: the case of the Netherlands, 1950–2009', paper presented at the I–O Society Conference, Sydney, 20–25 June.

Huppes, G., de Koning, A., Suh, S., Heijungs, R., Oers, L., van Nielsen, P. and Guinee, J. (2005) 'Environmental impacts of consumption in the EU: high resolution input ouput tables with detailed environmental extensions', *Journal of Industrial Ecology*, 10: 129–46.

IPTS (2006) 'Environmental Impacts of Products (EIPRO)', IPTS/ESTO project IPTS, European Commission technical report series, Seville: IPTS.

Jacobsen, H.K. (2000) 'Energy demand, structural change and trade: a decomposition analysis of the Danish manufacturing industry', *Economic Systems Research*, 12: 319–343.

Kagawa, S. and Inamura, H. (2001) 'A structural decomposition of energy consumption based on a hybrid rectangular input–output framework: Japan's case', *Economic Systems Research*, 13: 339–363.

Kagawa, S. and Inamura, H. (2004) 'A spatial structural decomposition analysis of Chinese and Japanese energy demand: 1985–1990', *Economic Systems Research*, 16: 279–299.

Lenzen, M. (2011) 'Aggregation versus disaggregation in input–output analysis of the environment', *Economic Systems Research*, 23: 73–89.

Marin, G. and Mazzanti, M. (in press), 'The evolution of environmental and labor productivity dynamics: sector analyses and decoupling/recoupling trends on a 1990–2007 Italian NAMEA', *Journal of Evolutionary Economics*, in press.

Mazzanti, M. and Montini, A. (2010) 'Embedding emission efficiency at regional level: analyses using NAMEA', *Ecological Economics*, 69(12): 2457–2467.

Miller, R.E. and Blair, P.D. (2009) *Input–Output Analysis: Foundations and Extensions*, 2nd edn, Cambridge: Cambridge University Press.

Moll, S., Vrgoc, M., Watson, D., Femia, A., Pedersen, O.G. and Villanueva, A. (2007) 'Environmental input–output analyses based on NAMEA data: a comparative European study on environmental pressures arising from consumption and production patterns', ETC/RWM Working Paper 2/2007, Copenhagen: European Topic Centre – Resource and Waste Management.

Morimoto, Y. (1970) 'On aggregation problems in input–output analysis', *Review of Economic Studies*, 37: 119–126.

Nakano, S., Okamura, A., Sukurai, N., Suzuki, M., Tojo, Y. and Yamano, N. (2009) 'The measurement of CO_2 embodiments in international trade: evidence from the harmonised input–output and bilateral trade database', Statistical Analysis of Science, Technology and Industry Working Paper 2009/3, Paris: OECD.

Peters, G. (2007) 'From production-based to consumption-based national emission inventories', *Ecological Economics*, 65: 13–23.

Roca, J. and Serrano, M. (2007) 'Income growth and atmospheric pollution in Spain: an input–output approach', *Ecological Economics*, 63: 230–242.

Rormose, P., Olsen, T. and Hansen, D. (2010) 'Emissions embodied in Danish import: a unidirectional trade model with 51 countries', paper presented at the I–O Society Conference, Sydney, 20–25 June.

Rose, A. and Casler, S. (1996) 'Input–output structural decomposition analysis: a critical appraisal', *Economic Systems Research*, 8: 33–62.

Rueda-Cantuche, J.M. (2010) 'Underestimation of the performance of the EU carbon dioxide emission reductions via external trade', paper presented at the I–O Society Conference, Sydney, 20–25 June.

Serrano, M. and Dietzenbacher, E. (2010) 'Responsibility and trade emission balances: an evaluation of approaches', *Ecological Economics*, 69: 2224–2232.

Stauvermann, P.J. (2007) 'The Regional Accounting Matrix including Environmental Accounts: methodology, application and interpretation – a Dutch pilot study'. Online: www.arpa.emr.it.

Su, B., Huang, H.C., Ang, B.W. and Zhou, P. (2010) 'Input–output analysis of CO_2 emissions embodied in trade: the effects of sector aggregation', *Energy Economics*, 32: 166–175.

Turner, K., Lenzen, M. and Wiedmann, J.B. (2007) 'Examining the global environmental impact of regional consumption activities – part 1: a technical note on combining input–output and ecological footprint analysis', *Ecological Economics*, 62: 37–44.

Watson, D. and Moll, S. (2008) 'Environmental benefits and disadvantages of economic specialisation within global markets, and implications for SCP monitoring', paper presented at the SCORE! conference, Brussels, Belgium, 10–11 March.

Wier, M. (1998) 'Sources of changes in emissions from energy: a structural decomposition analysis', *Economic Systems Research*, 10: 99–112.

Wyckoff, A.W. and Roop, J.M. (1994) 'The embodiment of carbon in imports of manufactured products: implications for international agreements on greenhouse gases emissions', *Energy Policy*, 22: 187–194.

9 NAMEA and the input–output framework

Sensitivity of environmental variables to changes in the production structure

Miguel Ángel Tarancón and Pablo del Río

Introduction

NAMEA is a methodology which allows integration of economic and environmental information in a coherent manner. The satellite accounts of the national accounting systems in European countries have been built using NAMEA methodology.

One of the advantages of using NAMEA to build environmental satellite account systems is that it offers the possibility of analysing the impact of productive activities at a disaggregated level per sector or per economic activity branch (de Haan and Keuning, 1996). In order to do so, the aforementioned environmental account systems should be linked to the national accounts dedicated to more disaggregated economic information, i.e. to the input–output framework.

The input–output framework is basically made up of supply-and-use tables, symmetric input–output tables and other complementary tables. These tables show the flow of goods and services between the different production activities and between these activities and the economic agents which make up the final demand of the economy for a given period and in monetary terms.

Input–output techniques can be used for a detailed and coherent analysis of the production flows of an economy and their environmental consequences. In this chapter, the symmetrical input–output table will be used since it allows us to analyse both the direct and indirect flows that stem from changes in final demand.

Accordingly, the chapter is structured as follows. The foundations of the input–output model applied to environmental issues are provided in next section, together with a classification of applications in the context of changes in the input–output coefficients. The third section goes deeper into the analysis of the variability of the coefficients and different techniques to analyse the sensitivity of NAMEA variables to changes in the productive structure of activity branches are reviewed. The advantages and drawbacks of the sensitivity analysis as well as several extensions to the environmental area are discussed in the final section.

Basic model and literature

There is an abundant literature linking the input–output framework to environmental issues and, thus, to environmental accounts. Leontief himself considered this possibility in the 1970s (Leontief, 1970) as a further development of his mathematical model, particularly through the inclusion of information on air pollutants (Leontief and Ford, 1972).

Before reviewing the different possibilities of using the input–output techniques in the environmental realm, the basic model is delineated.

Let us assume that the production activity of an economic system is made up of n productive sectors which interact in order to produce goods and services that meet final demand (private and public consumption, investment, net exports). Each productive sector consumes water and other natural resources and emits air pollutants and other residues. NAMEA allows us to identify the amount of resources, gases or residues which can be attributed to each production branch and to include them in the environmental satellite accounts. These substances, either consumed or generated, will be referred to as NAMEA substances from now on.

It will be assumed that the amount of the NAMEA substance consumed or generated e of the m sector can be broken down into the following two factors:

$$e_m = c_m \cdot x_m \tag{1}$$

where: c_m represents the intensity coefficient (i.e. substance per unit of output of sector m) and x_m is the quantity of output produced by m (in monetary terms). The level of x_m will depend on demand for the products of m by the rest of the sectors and by final demand:

$$x_m = \sum_{j=1}^{n} x_{mj} + y_m \tag{2}$$

where x_{mj} represent the sales of sector m to sector j, and y_m are the sales of m to the final demand.

On the other hand, the technical coefficients a_{mj} can be defined as the output of sector m required to produce a unit of product by sector j:

$$a_{mj} = \frac{x_{mj}}{x_j} \tag{3}$$

We know that activity j will have a direct requirement for the products of sector m but it will also have an indirect requirement for those products. This is due to the fact that the requirements of sector j for the products of a third sector k will lead to a requirement for the product of sector m by k, since k needs those (intermediate) products for its own production. This chain of additional production induced by sector m in order to produce the products of sector j may be extended indefinitely through the inclusion of other intermediate activities. The total requirements of a sector for the goods and services produced by the rest of the sectors can be picked up by the coefficients of the inverse Leontief matrix. Therefore, equation (2) can be rewritten as follows:

$$x_m = \sum_{q=1}^{n} b_{mq} \cdot y_q \tag{4}$$

where b_{mq} are the elements of the row m of the inverse Leontief matrix $B=(I-A)^{-1}$ which includes the direct and indirect purchases of the good or service produced by m made by sector q. A is the matrix $(n \times n)$ of technical coefficients which includes all the elements defined in (3).[1]

Furthermore, the vector of the coefficients of final demand can be defined per sector as:

$$h_i = \frac{y_i}{g} \tag{5}$$

where $g = \sum_{q=1}^{n} y_q$ represents the overall final demand which is an indicator of general economic activity. Equation (4) will then be transformed into:

$$x_m = \left(\sum_{q=1}^{n} b_{mq} \cdot h_q \right) \cdot g \tag{6}$$

Then,

$$e_m = c_m \cdot \left(\sum_{q=1}^{n} b_{mq} \cdot h_q \right) \cdot g \tag{7}$$

To sum up, three factors are behind the level of the NAMEA substance e of the m sector, including:

- *technological factors*, related to the technology used in sector m (supply-side): c_m;
- *scale factor* (demand side) which is determined by the level of economic activity (proxied by the overall final demand g);
- *structural factors*: these include the effects of the structure of final demand related to the share of each economic sector (represented by the final demand coefficients h_i). This final demand structure affects the level of the NAMEA substance, depending on the structure of the production functions of the productive activity branches which is determined by the technical coefficients a_{kl}.[2] These spread their effects on the level of NAMEA substance generation/consumption of sector m through the b_{mq} coefficients of the inverse Leontief matrix.

Once the basic model has been specified, we can distinguish three main lines of research which use the aforementioned framework in order to assess the environmental impacts of the economic system, with a disaggregation per sector.

A first group of authors have analysed the impact of changes in the final demand of the different sectors on either the global or the sectoral level of generation/consumption of a NAMEA substance through the analysis of demand multipliers (columns of coefficients of the Leontief matrix). This approach assumes structural stability of the productive relationships, i.e. the technical coefficients are considered fixed and only the 'flow' variables change.[3] If (7) is expressed in a disaggregated manner, then the following expression results:

$$\begin{aligned}\Delta e_m &= c_m \cdot \left(b_{m1}\Delta h_1 + b_{m2}\Delta h_2 + \ldots + b_{mn}\Delta h_n\right)\cdot \Delta g \\ &= c_m \cdot \left(b_{m1}\Delta y_1 + b_{m2}\Delta y_2 + \ldots + b_{mn}\Delta y_n\right)\end{aligned} \tag{8}$$

In other words, the impact of a sector on the NAMEA substance generation/consumption e due to changes in the final demand vector is assessed. In turn, these changes depend on variations in the level of economic activity (scale factor) and/or on changes in the expenditure and investment habits of actors (h_i coefficients). Several articles follow this approach, including: Proops (1988), Lenzen (1998), Munksgaard and Pedersen (2001), Machado *et al.* (2001), Ferng (2003), Sánchez-Chóliz and Duarte (2004), Gallego and Lenzen (2005), Mongelli *et al.* (2006), Peters and Hertwich (2006), Butnar and Llop (2007), Alcántara and Padilla (2009), Alcántara *et al.* (2010) and Chen and Zhang (2010).

A second set of studies within an input–output framework are based on structural decomposition techniques. The changes experienced by the level of generation/consumption of a NAMEA substance between two periods are explained by the changes in final demand and structural coefficients, within a comparative static approach:

$$\Delta e_m = c_m \cdot \left(\Delta b_{m1}\Delta h_1 + \Delta b_{m2}\Delta h_2 + \ldots + \Delta b_{mn}\Delta h_n\right)\cdot \Delta g \tag{9}$$

In contrast to the first set of studies, this approach analyses the impact of changes in the technical coefficients (7) on the level of the NAMEA polluting element under investigation. In other words, it involves a relaxation of the structural stability assumption. The drawback is that the distribution of the impacts between different factors (structure and demand flows) involves some arbitrariness and ad-hoc solutions are proposed (Dietzenbacher and Los, 1998; Rose and Chen, 1991; Chang and Lin, 1998; Wier, 1998; Kagawa and Inamura, 2001; de Haan, 2001; Hoen and Mulder, 2003; Alcántara and Duarte, 2004; Llop, 2007; Roca and Serrano, 2007; Tarancón and del Río, 2007c; and Liu *et al.*, 2010 belong to this stream of the literature).

A third group of approaches analyses the impact of changes in the technical coefficients in (7) on the level of the generation/consumption of a NAMEA substance from different activity branches. In contrast to the previous approach, the analysis does not focus on changes between two input–output tables in two different periods, but it simulates small variations taking place in the mathematical model which relates demand and NAMEA substance levels. The aim is to analyse the sensitivity of variations in the quantity of the NAMEA substance under investigation caused by those small changes. Therefore, it is worth looking at the following expression in more depth:

$$\Delta e_m = c_m \cdot \left(\Delta b_{m1} h_1 + \Delta b_{m2} h_2 + \ldots + \Delta b_{mn} h_n\right) \cdot g \tag{10}$$

Obviously, changes in the b_{mj} elements are the result of changes in the technical coefficients a_{ik} which represent the technological mix of each sector. Weber and Schnabl (1998), Tarancón and del Río (2007a, b) are relevant papers in this literature.

Analysis of the sensitivity of the coefficients

An analysis of the sensitivity of coefficients allows us to identify the transactions between activity branches which lead to the highest growth in the generation/consumption of a NAMEA substance. Accordingly, we can identify which production technologies (represented by the columns of technical coefficients) have the greatest impact on emissions.

Identification of structurally and technologically relevant coefficients

From equation (10) we can see that the quantity of e_m will depend on the production of sector m, x_m for a fixed factor c_m. For a level of fixed final demand (h_i fixed $\forall i$, and g fixed), the only variation in production will stem from a change in the coefficients of the inverse Leontief matrix (b) which belong to row m. In other words:

$$\Delta x_m = \left(\Delta b_{m1} h_1 + \Delta b_{m2} h_2 + \ldots + \Delta b_{mn} h_n\right) \cdot g \tag{11}$$

The changes in the b coefficients are obviously linked to changes in the technical coefficients of production which, if read per column, made up the technological mix of each sector. This relationship can be quantified through the Sherman and Morrison formula (see Sherman and Morrison, 1950). In particular, a change in the technical coefficient a_{ik} will induce the following changes in the elements of the inverse Leontief matrix b_{mj}:

$$\Delta b_{mj} = \frac{b_{mi} b_{kj} \Delta a_{ik}}{1 - b_{ki} \Delta a_{ik}} \tag{12}$$

Therefore, if (12) is substituted in (11), then:

$$\Delta x_m = \sum_{q=1}^{n} \frac{b_{mi} b_{kq} \Delta a_{ik}}{1 - b_{ki} \Delta a_{ik}} h_q g = \frac{b_{mi} x_k \Delta a_{ik}}{1 - b_{ki} \Delta a_{ik}} \tag{13}$$

since $\sum_{q=1}^{n} b_{kq} k_q g = x_k$.

Finally, this can be expressed as elasticities in this way:

$$\varepsilon_{a_{ik}x_m} = \frac{\dfrac{\Delta x_m}{x_m}}{\dfrac{\Delta a_{ik}}{a_{ik}}} = \frac{b_{mi}a_{ik}}{1-b_{ki}\Delta a_{ik}} \cdot \frac{x_k}{x_m} \tag{14}$$

Therefore, the elasticity of the generation/consumption of a NAMEA substance under investigation with respect to the change in the technical coefficient a_{ik} will be determined by the technological characteristics of the global production system and the output volume of activity branch k with respect to the output of sector m. This last factor depends on the demand for the products of sector k and, thus, it depends on final demand. Therefore, since this elasticity also depends on the composition of final demand in the economy, which determines the output of sector k, it does not only measure the relevance of the technological changes in the sector. Accordingly, we can call this the *structural* relevance factor since it is affected by the structure of the final demand vector.

On the other hand, we could identify the *technological relevance* of a coefficient (transaction) which is independent of the specific composition of the final demand vector. This is so because this composition does not depend on the firms in the activity branch but on the consumption and investment habits of economic actors and on the external sector (imports/exports).

To do this, we propose a uniform final demand vector as an alternative measure, i.e. $h_q g = 1 \forall q$. The changes in the generation/consumption of the NAMEA substance under investigation will result from the aggregation of the changes in the elements of row m of the Leontief matrix caused by the change in a_{ik}. Thus, the effect of the change of the row of the Leontief matrix can be expressed as follows:

$$\sum_{q=1}^{n} \Delta b_{mq} = \sum_{q=1}^{n} \frac{b_{mi}b_{kq}\Delta a_{ik}}{1-b_{ki}\Delta a_{ik}} \tag{15}$$

If the value reached by (15) is high, then the generation/consumption of a NAMEA substance under investigation will be highly sensitive to a change in the technological mix of sector k, independently of the structure of final demand. The only difference between (13) and (15) is that each of the b_{kq} coefficients is weighted by the final demand of sector q. (15) could be expressed as elasticities (relative increase in the aggregation of the elements of row m of matrix B due to a variation of the a_{ik} coefficient) in the following manner:

$$\varepsilon_{a_{ik}b_{m*}} = \frac{\dfrac{\Delta\sum_{q=1}^{n} b_{mq}}{\sum_{q=1}^{n} b_{mq}}}{\dfrac{\Delta a_{ik}}{a_{ik}}} = \frac{\dfrac{\sum_{q=1}^{n} \dfrac{b_{mi}b_{kq}\Delta a_{ik}}{1-b_{ki}\Delta a_{ik}}}{\sum_{q=1}^{n} b_{mq}}}{\dfrac{\Delta a_{ik}}{a_{ik}}} = \frac{\sum_{q=1}^{n} \dfrac{b_{mi}b_{kq}\Delta a_{ik}a_{ik}}{1-b_{ki}\Delta a_{ik}}}{\sum_{q=1}^{n} b_{mq}\Delta a_{ik}} = \frac{\sum_{q=1}^{n} \dfrac{b_{mi}b_{kq}a_{ik}}{1-b_{ki}\Delta a_{ik}}}{\sum_{q=1}^{n} b_{mq}} \tag{16}$$

Equation (16) provides information on the capacity that the technology of sector k has to induce the generation/consumption of the NAMEA substance in the whole production system when the coefficient a_{ik} changes, independently of the composition of final demand. In other words, (16) informs on the technological relevance of sector k due to a change in this coefficient.

Identification of relevant coefficients as a function of the global impact on the system

In the previous section, the relevance of the technical coefficients (and thus, the transactions between associated activity branches) has been identified depending on the impact that changes in the quantity of those technical coefficients have on the generation/consumption of a NAMEA substance by sector m. Alternatively, we could identify the most relevant coefficients depending on how their changes affect the generation/consumption of a NAMEA substance in the whole system, i.e. in all the sectors combined.

Accordingly, we propose an alternative measure for evaluating the relevance of a coefficient based on the concept of elasticity. This measure is based on a coefficient's capacity (elasticity) to change the overall output. This elasticity would be expressed as:

$$\varepsilon_{e_{\bullet}a_{ik}} = \frac{\dfrac{\sum_{m=1}^{n} \Delta e_m}{\sum_{m=1}^{n} e_m}}{\dfrac{\Delta a_{ik}}{a_{ik}}} \tag{17}$$

If we integrate (12) into (17) and take into account $x_k = \sum_{q=1}^{n} b_{kq}h_q g$, then:

$$\varepsilon_{e_{\bullet}a_{ik}} = \frac{\dfrac{\sum_{m=1}^{n} c_m \sum_{q=1}^{n} \dfrac{b_{mi}b_{kq}\Delta a_{ik}}{1-b_{ki}\Delta a_{ik}} h_q g}{\sum_{m=1}^{n} c_m x_m}}{\dfrac{\Delta a_{ik}}{a_{ik}}} = \frac{x_k a_{ik} \sum_{m=1}^{n} c_m b_{mi}}{\left(1-b_{ki}\Delta a_{ik}\right)\sum_{m=1}^{n} c_m x_m} \tag{18}$$

The most structurally important coefficients are those with the highest elasticity. Changes in those coefficients will generate the greatest variations in the overall amount of the generation/consumption of a NAMEA substance.

Discussion and conclusions

This chapter has highlighted the role of input–output analysis as a tool for exploiting the information provided by the NAMEA methodology. Several approaches allow us to use input–output tables to derive relevant economic and environmental results. These approaches fall broadly into two groups.

On the one hand, several techniques are based on the structural stability hypothesis and, thus, aim to assess the consequences of changes in the flow variables of the system (especially, components of final demand by activity branches) for a given technological structure.

On the other hand, other techniques focus on the changes in the production structure of the system. This structure materializes in the matrix of technical coefficients for the input–output system.

In this second group, the chapter has aimed to provide an overview of sensitivity analyses in input–output techniques applied to the generation/consumption of the NAMEA substance. Sensitivity analyses allow us to assess the environmental impact of the different coefficients which make up the productive structure of the economic system. They allow us to obtain a map of the *hot-points* of the production system, i.e. to identify the transactions between sectors which have the greatest impact on the generation/consumption of the NAMEA substance. This approach is deemed highly relevant in the context of environmental impact assessment since it is able to identify the degree of importance of a sector in the generation of emissions by taking into account both direct and indirect impacts.

However, at least four limitations in this approach are worth mentioning.

First, the input–output approach assumes linearity of productive relationships (i.e. constant returns to scale), which is not very realistic. Nonetheless, this assumption does not significantly distort the results when the focus of the analysis is small structural changes.

Second, the sensitivity analysis assumes that the rest of technical coefficients remain constant. This *ceteris paribus* assumption is very restrictive since it rules out the impact of joint changes in the coefficients. Fortunately, empirical studies have shown that the impact of those joint effects is of minor importance compared with the individual effects.

Third, the quantity input–output model considers the final demand vector as an exogenous variable. The consideration of the distribution coefficients of final demand (h_i) allows us to analyse the structure of final demand. However, this does not imply that the scale factor (g) changes in an endogenous manner. One alternative to make those effects endogenous is to use the modelling framework of the social accounting matrix (SAM) and to build models in which the structure of final demand is endogenous within an extended inverse Leontief matrix.

Examples of this alternative are Manresa and Sancho (2004), Sánchez-Chóliz *et al.* (2007), Hartono and Resosudarmo (2008), Parikh *et al.* (2009) and Duarte *et al.* (2010) who use a SAM to make private consumption and their impact on greenhouse emissions endogenous. In turn, Rodríguez-Morilla *et al.* (2007) use the matrix of multipliers of a SAM in order to analyse the economic and social efficiency of the productive system.

Fourth, the aforementioned techniques use quantities produced. In contrast, the impact of prices on the generation/consumption of the NAMEA substance is not explicitly considered. Although there is a price version in the input–output model (see, among others, Førsund, 1985; Llop and Pié, 2008; Tarancón *et al.* 2010a), a joint treatment of quantities and prices requires a more flexible modelling framework. In this context, the literature has advanced through the construction of general computable equilibrium (CGE) models which use information from the previously built SAM. Examples of these models include Li and Rose (1995), Naqvi (1998), Zhang (1998), Allan *et al.* (2007), Guivarch *et al.* (2009) and Kretschmer and Peterson (2010).

Finally, we should go beyond the analysis of the interactions between the activity branches of an economic system, as in this chapter. The interactions between the activity branches of different economic systems should also be analysed. One stream of the literature on input–output analysis has gone deeply into these aspects through the creation of multi-regional input–output models (MRIO) which require information on international trade flows. For example, Reinert and Roland-Holst (2001) build a model based on the SAM for Canada, Mexico and the United States. Other environmental and multi-regional applications of the input–output framework are Liang *et al.* (2006), Ackerman *et al.* (2007), Li and Hewitt (2008) and McGregor *et al.* (2008). A detailed review of these models is provided by Wiedmann *et al.* (2007) and Wiedmann (2009).

Notes

1 The Leontief model can be expressed as follows. Let x be the vector (nx1) of sector production, A the matrix of technical coefficients defined in (3) and y the vector (nx1) of final demand. Therefore: $x=Ax+y \Rightarrow x-Ax=y \Rightarrow (I-A)x=y \Rightarrow x=(I-A)^{-1}y$, with $B=(I-A)^{-1}$ being the Leontief inverse matrix.
2 The terms sector and branch are used interchangeably.
3 Several applications use this approach in order to assess the impacts of foreign trade on energy consumption and CO_2 emissions.

References

Ackerman, F., Ishikawa, M. and Suga, M. (2007) 'The carbon content of Japan–US trade', *Energy Policy*, 35: 4455–4462.

Alcántara, V. and Duarte, R. (2004) 'Comparison of energy intensities in European Union countries: results of a structural decomposition analysis', *Energy Policy*, 32: 177–189.

Alcántara, V. and Padilla, E. (2009) 'Input–output subsystems and pollution: an application to the service sector and CO_2 emissions in Spain', *Ecological Economics*, 6: 905–914.

Alcántara, V., del Río, P. and Hernández, F. (2010) 'Structural analysis of electricity consumption by productive sectors: the Spanish case', *Energy*, 35: 2088–2098.

Allan, G., Hanley, N., McGregor, P., Swales, K. and Turner, K. (2007) 'The impact of increased efficiency in the industrial use of energy: a computable general equilibrium analysis for the United Kingdom', *Energy Economics*, 29: 779–798.

Butnar, I. and Llop, M. (2007) 'Composition of greenhouse gas emissions in Spain: an input–output analysis', *Ecological Economics*, 61: 388–395.

Chang, Y.F. and Lin, S.J. (1998) 'Structural decomposition of industrial CO_2 emissions in Taiwan: an input–output approach', *Energy Policy*, 26: 5–12.

Chen, G.Q. and Zhang, B. (2010) 'Greenhouse gas emissions in China 2007: inventory and input–output analysis', *Energy Policy*, 38: 6180–6193.

de Haan, M. (2001) 'A structural decomposition analysis of pollution in the Netherlands', *Economic Systems Research*, 13: 181–196.

de Haan, M. and Keuning, S.J. (1996) 'Taking the environment into account: the NAMEA approach', *Review of Income and Wealth*, 42: 131–148.

Dietzenbacher, E. and Los, B. (1998) 'Structural decomposition techniques: sense and sensitivity', *Economic Systems Research*, 4: 307–323.

Duarte, R., Mainar, A. and Sánchez-Chóliz, J. (2010) 'The impact of household consumption patterns on emissions in Spain', *Energy Economics*, 32: 176–185.

Ferng, J.J. (2003) 'Allocating the responsibility of CO_2 over-emissions from the perspectives of benefit principle and ecological deficit', *Ecological Economics*, 46: 121–141.

Fontela, E. (2005) 'Conferencia inaugural', Spanish Input–Output Conference, University of Oviedo, mimeo.

Førsund, F.R. (1985) 'Input–output models, national economic models, and the environment', in A.V. Kneese and J.L. Sweeney (eds), *Handbook of Natural Resource and Energy Economics*, Amsterdam: Elsevier, pp. 325–344.

Gallego, B. and Lenzen, M. (2005) 'A consistent input–output formulation of shared producer and consumer responsibility', *Economic Systems Research*, 17: 365–391.

Guivarch, C., Hallegatte, S. and Crassous, R. (2009) 'The resilience of the Indian economy to rising oil prices as a validation test for a global energy–environment–economy CGE model', *Energy Policy*, 37: 4259–4266.

Hartono, D. and Resosudarmo, B.P. (2008) 'The economy-wide impact of controlling energy consumption in Indonesia: an analysis using a Social Accounting Matrix framework', *Energy Policy*, 36: 1404–1419.

Hoen, A. and Mulder, M. (2003) 'Explaining Dutch emissions of CO_2: a decomposition analysis', CPB discussion paper, The Hague: Netherlands Bureau for Economic Policy Analysis.

Kagawa, S. and Inamura, H. (2001) 'A structural decomposition of energy consumption based on a hybrid rectangular input–output framework: Japan's case', *Economic Systems Research*, 13: 339–363.

Kretschmer, B. and Peterson, S. (2010) 'Integrating bioenergy into computable general equilibrium models: a survey', *Energy Economics*, 32: 673–686.

Lenzen, M. (1998) 'Primary energy and greenhouse gases embodied in Australian final consumption: an input–output analysis', *Energy Policy*, 26: 496–506.

Leontief, W. (1970) 'Environmental repercussions and the economic structure: an input–output approach', *Review of Economics and Statistics*, 52: 262–271.

Leontief, W. and Ford, D. (1972) 'Air pollution and the economic structure: empirical results of input–output computations', in A. Brody and A. Carter (eds), *Input–Output Techniques*, Amsterdam: North-Holland.

Li, P. and Rose, A. (1995) 'Global warming policy and the Pennsylvania economy: a computable general equilibrium analysis', *Economic Systems Research*, 7: 151–171.

Li, Y. and Hewitt, C.N. (2008) 'The effect of trade between China and the UK on national and global carbon dioxide emission', *Energy Policy*, 36: 1907–1914.

Liang, Q.M., Fan, Y. and Wei, Y.M. (2006) 'Multi-regional input–output model for regional energy requirements and CO_2 emissions in China', *Energy Policy*, 35: 1685–1700.

Liu, H., Xi, Y., Guo, J. and Li, X. (2010) 'Energy embodied in the international trade of China: an energy input–output analysis', *Energy Policy*, 38: 3957–3964.

Llop, M. (2007) 'Economic structure and pollution intensity within the environmental input–output framework', *Energy policy*, 35: 3410–3417.

Llop, M. and Pié, L. (2008) 'Input–output analysis of alternative policies implemented on the energy activities: an application for Catalonia', *Energy Policy*, 36: 1642–1648.

McGregor, P.G., Swales, J.K. and Turner, K. (2008) 'The CO_2 trade balance between Scotland and the rest of the UK: performing a multi-region environmental input–output analysis with limited data', *Ecological Economics*, 66: 662–673.

Machado, G., Schaeffer, R. and Worrell, E. (2001) 'Energy and carbon embodied in the International Trade of Brazil: an input–output approach', *Ecological Economics*, 39: 409–424.

Manresa, A. and Sancho, F. (2004) 'Energy intensive and CO_2 emissions in Catalonia: a SAM analysis', *International Journal of Environment, Workforce and Employment*, 1: 91–106.

Mongelli, I., Tassielli, G. and Notarnicola, B. (2006) 'Global warming agreements, international trade and energy/carbon embodiments: an input–output approach to the Italian case', *Energy Policy*, 34: 88–100.

Munksgaard, J. and Pedersen, K.A. (2001) 'CO_2 accounts for open economies: producer or consumer responsibility', *Energy Policy*, 29: 327–334.

Naqvi, F. (1998) 'A computable general equilibrium model of energy, economy and equity interactions in Pakistan', *Energy Economics*, 20 (4): 347–373.

Parikh, J., Panda, M., Ganesh-Kumar, A. and Singh, V. (2009) 'CO_2 emissions structure of Indian economy', *Energy*, 34: 1024–1031.

Peters, G.P. and Hertwich, E.G. (2006) 'Pollution embodied in trade: the Norwegian case', *Global Environmental Change*, 16: 379–387.

Proops, J.L.R. (1988) 'Energy intensities, input–output analysis and economic development', in M. Ciaschini (ed.), *Input–Output Analysis: Current Developments*, London: Chapman and Hall.

Reinert, K.A. and Roland-Holst, D.W. (2001) 'Industrial pollution linkages in North America: a linear analysis', *Economic Systems Research*, 13: 197–208.

Roca, J. and Serrano, M. (2007) 'Income growth and atmospheric pollution in Spain: an input–output approach', *Ecological Economics*, 63: 230–242.

Rodríguez-Morilla, C., Llanes, G. and Cardenete, M.A. (2007) 'Economic and environmental efficiency using a social accounting matrix', *Ecological Economics*, 60: 774–786.

Rose, A. and Chen, C.Y. (1991) 'Sources of change in energy use in the U.S. economy, 1972–1982', *Resources and Energy Economics*, 13: 1–21.

Sánchez-Chóliz, J. and Duarte, R. (2004) 'CO_2 emissions embodied in international trade: evidence for Spain', *Energy Policy*, 32: 1999–2005.

Sánchez-Chóliz, J., Duarte, R. and Mainar, A. (2007) 'Environmental impact of household activity in Spain', *Ecological Economics*, 62: 308–318.

Sherman, J. and Morrison, W.J. (1950) 'Adjustment of an inverse matrix corresponding to a change in one element of a given matrix', *Annals of Mathematical Statistics*, 21: 124–127.

Tarancón Morán, M.A. and Del Río González, P. (2007a) 'CO_2 emissions and intersectoral linkages: the case of Spain', *Energy Policy*, 35: 1100–1116.

Tarancón Morán, M.A. and Del Río González, P. (2007b) 'A combined input–output and sensitivity analysis approach to analyse sector linkages and CO_2 emissions', *Energy Economics*, 29.

Tarancón Morán, M.A. and Del Río González, P. (2007c) 'Structural factors affecting land-transport CO_2 emissions: a European comparison', *Transportation Research Part D: Transport and Environment*, 12.

Tarancón Morán, M.A., Callejas Albiñana, F., Dietzenbacher, E. and Lahr, M.L. (2008a) 'A revision of the tolerable limits approach: searching for the important coefficients', *Economic Systems Research*, 20.

Tarancón Morán, M.A., Del Río González, P. and Callejas Albiñana, F. (2008b) 'Tracking the genealogy of CO_2 emissions in the electricity sector: an intersectoral approach applied to the Spanish case', *Energy Policy*, 36: 1915–1926.

Tarancón Morán, M.A., Del Río González, P. and Callejas Albiñana, F. (2010) 'Assessing the influence of manufacturing sectors on electricity demand: a cross-country input–output approach', *Energy Policy*, 38: 1900–1908.

Tarancón Morán, M.A., Del Río González, P. and Callejas Albiñana, F. (in press) 'Determining the responsibility of manufacturing sectors regarding electricity consumption: the Spanish case', *Energy*.

Weber, C. and Schnabl, H. (1998) 'Environmentally important intersectoral flows: insights from main contributions identification and minimal flow analysis', *Economic Systems Research*, 10: 337–356.

Wiedmann, T. (2009) 'A review of recent multi-region input–output models used for consumption-based emission and resource accounting', *Ecological Economics*, 69: 211–222.

Wiedmann, T., Lenzen, B., Turner, K. and Barretta, J. (2007) 'Examining the global environmental impact of regional consumption activities – part 2: review of input–output models for the assessment of environmental impacts embodied in trade', *Ecological Economics*, 61: 15–26.

Wier, M. (1998) 'Sources of changes in emissions from energy: a structural decomposition analysis', *Economic Systems Research*, 10: 99–112.

Zhang, Z. (1998) 'Macro-economic and sectoral effects of carbon taxes: a general equilibrium analysis for China', *Economic Systems Research*, 10: 135–159.

10 Environmental impact of generating electricity by replacing lignite with photovoltaic technology

An analysis on the Greek NAMEA

Anastasia Basina, Charalambos Economidis and Athanasios Sfetsos

Introduction

The greenhouse effect has led to a number of environmental problems such as an increase in the average temperature of the Earth. The main cause of this effect is high CO_2 emissions (European Environment Agency, 2010). The measures that are taken to deal with air pollution may be technical and legislative/economic. The main technical measures involve the construction of new power plants with low-intensity emissions and the application of technologies with increased energy efficiency.

The Kyoto Protocol also aims to reduce greenhouse gases, mainly those emitted by developed countries. The ways of achieving the targets are primarily the introduction of national policies to reduce emissions and cooperation among the signatory states. In Europe, the protocol envisages an 8 per cent reduction in greenhouse gas emissions by the signatory states between 2008 and 2012 (below 1990 levels). Greece was allowed an increase in its emissions of 25 per cent by 2012 compared with 1990. By 2012, however, emissions are expected to have risen by more than 25 per cent, which means Greece will not meet the Kyoto target unless it takes immediate measures (European Environment Agency, 2010).

The problem can also be tackled by replacing conventional energy sources with renewable ones. Renewable energy sources do not run out but are constantly replenished and can be easily converted into electrical or thermal energy. Examples are solar energy, photovoltaic systems and solar thermal systems.

The implementation in Greece of Law 3468/2006 concerning the generation of electricity using renewable energy sources made it much simpler to install photovoltaic systems.[1] The target set by the Ministry of Development was the installation of 200 MW of photovoltaic energy-generation systems in Greece by the end of 2010.[2]

In this section, calculations have been made in relation to Greece so that the country can avoid sanctions arising from its failure to achieve the targets set in the Kyoto Protocol. Greece must reduce by 8 per cent its CO_2 emissions which cause the greenhouse effect.

On the basis of data taken from the PPC's operational programme, certain calculations were made for a reduction in CO_2 emissions. The installed capacity must be reduced and the total lignite used will have to be reduced by 8 per cent compared with 2005. Lastly, demand for electricity is assumed to rise or at least remain constant at 2005 levels. Consequently, this reduction in net electricity production resulting from the decrease in the amount of lignite will have to be offset by electrical energy generated from the installation of photovoltaic systems. It has been assumed that the amount of electrical energy which must be generated by photovoltaic system technology is equal to the size of the reduction in electricity generation from the combustion of lignite. Moreover, the amount of energy required from the installation of photovoltaic technology was also calculated.

According to the 2005 NAMEA table, CO_2 is the main factor responsible for air pollution and its amount was calculated for the electrical energy industry. CO_2 emissions for the electrical energy industry after the 8 per cent reduction were also calculated. This change in turn led to a change in the direct coefficients of intensity of the Global Warming Potential (GWP) group. The change in the direct coefficients of air emission intensity results in a change in the direct and indirect impact data and an extension in the elasticity of emission intensity in relation to production and final demand. In this way, the total environmental impact (direct and indirect) is calculated for the electrical energy industry. Lastly, the direct and indirect emission intensity per unit of final demand categories was computed. The change in final demand per unit for the electrical energy industry initially produced a coefficient for direct and indirect emissions whereas after our assumption (8 per cent reduction in CO_2 emissions), this coefficient changes.

Definition of NAMEA

According to Eurostat (2001), the National Accounting Matrix including Environmental Accounts (NAMEA) constitutes a framework in which various types of statistical data have been consistently organized in a way that brings together economic and environmental information that originates from different parts of the statistical system. The core of this framework is a set of tables containing economic data and forms a National Accounting Matrix (NAM) compiled for National Accounts. The Environmental Accounts (EA) consists of tables containing data that are often denominated in physical units. The types of physical environmental data taken into consideration in the NAMEA accounting framework include air emissions in particular, but also emissions of waste, wastewater and water use. The Energy Balance shows the supply and use of energy and stock data on subsoil and in forest assets. The NAM also enables the presentation of environmental taxes, expenditures and other economic data.

Figure 10.1 shows how National Accounting Tables of Supply and Use and Environmental Accounts (emissions) can be expanded in the context of a NAMEA. Here, the NAM is made up of the supply-and-use tables of ordinary National Accounts, to which household consumption is added according to

<table>
<tr><td colspan="4">Ordinary National Account</td><td>Physical data</td></tr>
<tr><td></td><td colspan="3">Supply</td><td></td></tr>
<tr><td rowspan="3">Use</td><td>Trade margins</td><td>Intermediate consumption</td><td>Final uses</td><td>Household emissions</td></tr>
<tr><td>Output</td><td></td><td></td><td>Emissions from industries and services</td></tr>
<tr><td>Imports</td><td>Value added</td><td></td><td></td></tr>
<tr><td></td><td colspan="4">Employment energy and physical data</td></tr>
</table>

Figure 10.1 Schematic description of a NAMEA framework for air emissions (source: Eurostat, 2001).

purpose. As can be seen on the right side of the figure, the production activities of industries result in emissions as well as in goods and services that are included in regular National Accounts. Household consumption and the corresponding emissions are displayed separately. Below the supply-and-use tables, the industry use of labour, energy and other materials can also be presented in physical units. Of course the NAM can also include other National Account Matrices, for example showing income, saving and financial flows. As an alternative, the NAMEA could include input–output tables instead of supply and use matrices.

A NAMEA pilot study for Greece

The 'Natural Resource Accounts and Environmental Input–Output Tables 1988–1998' report (Mylonas *et al.*, 2000) was the first systematic attempt to develop a consistent NAMEA for Greece. In this study, the environmental accounts showing air emissions (which include energy and material balances, accounts for water, wood and fish) cover the period 1988–1996. The air emission accounts take into consideration seven pollutants (CO_2, N_2O, CH_4, SO_2, NO_x, CO and NMVOC) included in CORINAIR. The emissions are shown in accounts in accordance with a classification that is comparable with NACE Rev. 1. In the study (Mylonas *et al.*, 2000), the calculation of the air emission data is based on CORINAIR data available according to the Selected Nomenclature for Sources of Air Pollution (SNAP) 1994. They are allocated to a 25 industry-level classification. Household emissions are taken into account since 1990 whereas the relation between transport and heating is not shown. The most recent NAMEA table constructed for Greece is for 2005 (Economidis *et al.*, 2008).

Table 10.1 NACE* activities according to the 2005 Greek input–output table (26 industries)

Code	*NACE Activity Rev. 1*
01 & 02	Agriculture
5	Fisheries
10, 11 & 12/13–14	Mining and quarrying
15–16	Manufacture of food products, beverages and tobacco
17–19	Manufacture of textiles and textile products
20A	Manufacture of wood and wood products
21–22	Manufacture of pulp, paper and paper products; publishing and printing
23	Manufacture of coke, refined petroleum products and nuclear fuel
24–25	Manufacture of chemicals, chemical products and man-made fibres
26	Manufacture of other non-metallic mineral products
27	Manufacture of basic metals and fabricated metal products
28	Manufacture of fabricated metal products, except machinery and equipment
29–36	Manufacture of machinery and equipment
37	Recycling
40–41	Electricity, gas and water supply
45	Construction
50–52	Wholesale and retail trade; repair of motor vehicles, motorcycles and personal and household goods
55	Hotels and restaurants
60–64	Transport, storage and communication
65–67	Financial intermediation
70–74	Real estate, renting and business activities
75 & 90	Public administration and defence; Sewage and refuse disposal
80–84	Education
85	Health and social work
91	Activities of membership organizations n.e.c.
92, 93, 95 & 99	Recreational, cultural and sporting activities; Activities of households; Extra-territorial organizations

Note
*Nomenclature statistique des activités économiques dans la communauté européenne.

Environmental data used for this study

According to Economidis *et al.* (2008), the air emission data are estimated using the so-called 'air emission inventory first approach' which is mainly based on the allocation of energy balance data to NACE categories (Eurostat, 2007). At present, international agreements on air emissions include the CLRTAP (Convention on Long-Range Transboundary Air Pollution) with reporting to UNECE/EMEP and the UNFCCC (United Nations Framework Convention on Climate Change), and reporting based on the UNFCCC CRF (Common Reporting Format). The UNFCCC CRF covers six categories of greenhouse gases (CO_2, N_2O, CH_4, HFCs, PFCs and SF_6) plus four indirect greenhouse gases (NO_x, CO, NMVOC, SO_2). UNECE/EMEP reporting includes only NO_x, CO, NMVOC and SO_2 plus NH_3 plus nine heavy metals as well as 17 POPs (persistent organic pollutants).

The air emissions have been further grouped and aggregated by three environmental pressure variables, namely Global Warming Potential (GWP), Acidification (ACID) and Tropospheric Ozone Forming Potential (TOFP) in addition to particulate matter (PM_{10}) with a diameter of less than 10 mm, from the following set of equations:

$$\text{GWP} = \text{CO}_2 + 310 * \text{N}_2\text{O} = 21 * \text{CH}_4 \quad (1)$$

$$\text{ACID} = \text{SO}_2 + 0.7 * \text{NO}_x + 1.9 * \text{NH}_3 \quad (2)$$

$$\text{TOFP} = \text{NMVOC} + 1.22 * \text{NO}_x + 0.11 * \text{CO} + 0.014 * \text{CH}_4 \quad (3)$$

The proposed grouping results from the combined effects that each primary pollutant has on ecosystems. GWP is a measure of climate change whereas ACID is primarily related to the effect of acid rain, eutrophication of ecosystems and cardiovascular diseases in humans. TOFP is related to the production of ozone in the lowest atmosphere and can be a major factor influencing the respiratory (e.g. asthma) and cardiovascular systems of humans. Both TOFP and ACID have been found to severely damage monuments of significant cultural heritage. PM_{10} is presently one of the most important pollutants associated with respiratory problems and carcinogenic potency. In a study conducted on 29 European cities (Katsouyanni *et al.*, 2001), the result for the city of Athens showed that an increase in daily PM_{10} levels by 10 mg/m^3 results in an increase in the mortality rate of approximately 1.5 per cent.

Calculation of eco-indicators

Methodology

The pollution indicators have been constructed using the following methodology (Economidis *et al.*, 2008):

We can express the input–output model in matrix terms:

$$\mathbf{X} = \mathbf{AX} + \mathbf{Y} \ldots \tag{4}$$

Matrix **A** is known as the matrix of technical coefficients, **X** the vector of total resources of the industry and **Y** the vector of final use of products. Solving the equation (4) for **X**, we obtain:

$$\mathbf{X} = (\mathbf{I} - \mathbf{A})^{-1}\mathbf{Y} \ldots \tag{5}$$

where $(\mathbf{I}-\mathbf{A})^{-1}$ is often referred to as the Leontief inverse. We applied a domestic-oriented model, starting from final demand and quantifying the direct and indirect effects on production and emissions activated by final demand. The results can be presented as a re-attribution of the domestic production which is derived from the basic Leontief model:

$$(\mathbf{I} - \mathbf{A}^{\mathrm{d}})\mathbf{X} = (\mathbf{F}^{t} - \mathbf{M}) \ldots \tag{6}$$

$\mathbf{A}^{\mathrm{d}}$ matrix of domestic technical coefficients ($\mathbf{A}^{t} = \mathbf{A}^{d} + \mathbf{A}^{m}$),
$\mathbf{F}^{t}$ vector of total (domestic + imported) final demand by industry,
M the import vector by industry which is subtracted from total final demand.

The direct emission intensity coefficients a_{kj} of the k type of emission by industry j are defined by:

$$a_{kj} = E_{kj}/X_j \quad j = 1, \ldots, n \ldots \tag{7}$$

E_{kj} quantity of emission k by industry j (components of vector **E**),
X_j domestic value of output of industry j.

The coefficients a_{kj} show the extent to which each industry generates a certain direct emission intensity factor and describe the physical quantity of emission directly caused by the production of goods as a ratio of the value of domestic production at basic prices. We also compute the total environmental dependencies:

$$\varepsilon_{kj} = \hat{\mathbf{a}}_{kj}(\mathbf{I} - \mathbf{A}^{\mathrm{d}})^{-1} \ldots \tag{8}$$

where ε_{kj} is the matrix of induced emission coefficients showing the emission intensity, including the induced emission in other industries indirectly affected by a change in final demand of a given industry (Mylonas *et al*., 2000, p. 60).

The elasticity of emission intensity with respect to final consumption e_j^c is estimated as the row sum of ε_{kj}, $e_j^c = \sum_j \varepsilon_{kj}$, including direct and indirect emissions for the *j-th* industry. These emissions are produced in the whole economy after an increase in a unitary expansion of the final demand of the specific industry.

The column sum of ε_{kj}, $e_k^p = \sum_k \varepsilon_{kj}$ is an estimate of the total emissions of the pollutants in the whole economy owing to a one-unit increase in a given industry's production, termed as the elasticity of emission intensity with respect to production.

The direct and indirect emission intensity per unit of final demand categories is computed by the following operation:

$$\varepsilon^d = \hat{\varepsilon}_{kj} F^t \ldots \tag{9}$$

where ε^d is the industry's extent of each emission factor intensity on each final demand component.

Electricity generation mix and CO_2 emissions

The electricity generation mix has been clear since PPC published its annual programme for 2005. As can be seen, lignite is the main factor of production in the generation of electrical energy. A total of 60.41 per cent of electrical energy comes from the conventional combustion of lignite, which is a basic commodity in Greece, since it is in abundance and its price is very low compared with other energy sources. For this reason, and despite the fact that its combustion causes the most emissions, lignite continues to be used for the generation of electricity. In 2005, natural gas and oil were used in more or less the same proportions of around 15 per cent each. It is worth noting that natural gas produces half the emissions from generating 1 kWh of electrical energy compared with emissions from the combustion of lignite. The remaining 10 per cent of electrical energy comes from hydroelectric installations and only 0.14 per cent from Renewable Energy Sources (RES). The percentage of electrical energy generated by photovoltaic systems is very low although efforts are being made to use them and the state offers grants and other incentives to both industry and households to encourage them to use photovoltaic systems for the generation of electrical energy (in small- and large-scale installations).

Table 10.4 shows emissions and how much they will be reduced (in g) if conventional energy sources are replaced by photovoltaic systems. More specifically, the reduction in the quantity of emissions when 1 kWh is produced by PV systems.

Table 10.2 Electricity generation mix for 2005 (%)

PPC electricity generation mix for 2005	
Lignite	60.41
Oil	14.85
Natural gas	14.46
Hydroelectric	10.14
RES (renewable energy sources)	0.14

Source: Greenpeace (2006).

Table 10.3 Net generation and corresponding CO_2 emissions for 2005

Fuel	*Net generation (GWh)*	*Emissions of CO_2 (kton)*
Lignite	31,977	43,041
Oil (in the transmission grid system)	3,300	2,566
Oil (islands)	4,566	3,529
Natural gas	7,654	3,509
Hydroelectric	5,366	0
Renewable Energy Sources (PPC)	74	0
Total	52,937	52,645

Source: Kanellopoulos *et al.* (2007).

Table 10.4 Reduction of emissions through the replacement of conventional energy sources with photovoltaic systems

Substitute	*Reduction of emissions (in g) per solar kilowatt-hour*			
	CO_2	*SO_2*	*NO_x*	*PM_{10}*
Lignite	1,482	1–1.8	1.17–1.23	1.1
Oil	830	3.5	1.5	0.34
(Low sulphur)				
Natural gas	475	0.017	0.6	
Country's average energy mix	1,062			

Source: Environment, Education, Sustainability (2007).

Notes

CO_2: carbon dioxide; SO_2: sulphur dioxide; NO_x: nitrogen oxide; PM_{10}: particulate matter.

It can be seen in Table 10.4 that by replacing the method of producing 1 kWh (using photovoltaic systems instead of lignite, we will see a reduction of 1,482 g CO_2, 1–1.8 g SO_2, 1.17–1.23 g NO_x, 1.1 g PM_{10}).

The change in the electricity generation mix and the replacement of conventional energy source with photovoltaic systems is necessary in order to achieve the targets set by the Kyoto Protocol.

The use of RES for the generation of electricity is increasing. The most common RES in Greece is wind power, although due to its geographical position, the country also has potential for the development of installations producing solar energy and hydropower. Between 2005 and 2008, there has been a relative growth in RES overall of around 46.5 per cent, while with regard to photovoltaic systems, the installed capacity rose from 0.51 MW in 2005 to 0.68 MW (+25 per cent) in 2006 and 0.74 MW (+8 per cent) in 2007. However, the biggest increase was seen in 2008, with an installed photovoltaic capacity of 10.49 MW (+92.9 per cent). This indicates that Greece is responding positively to the potential offered by RES.

Table 10.5 Installed capacity in MW from RES

Year RES	*2005*	*2006*	*2007*	*2008*
Wind	576.09	749.27	853.62	997.56
Biomass	20.54	37.58	37.57	39.40
Hydroelectric	48.16	73.68	95.5	150.88
Photovoltaic	0.51	0.68	0.74	10.49
Total	645.30	861.21	987.43	1,198.33

Source: Monthly Bulletin, Hellenic Transmission System Operator, November (2008).

Results of the completion of PPC's strategic plan vis-à-vis the environment

The successful completion of PPC's strategic plan will result in significant mitigation of the impact on the environment from the generation of electrical energy in Greece, while at the same time maintaining an even mix of fuels in the energy balance. With the replacement of old production plants and the operation of new, modern power stations (fuelled by lignite, coal, natural gas and hydropower) and the increased penetration of renewable energy sources for the production of 1 kWh in the transmission grid system, emissions into the environment in 2015 will be much lower than present levels.

At the same time, the completion of the strategy (PPC, 2008) will significantly contribute towards the fulfilment of obligations undertaken by Greece, 'since these obligations will be shaped by the targets of the EU's Thematic Strategy on Air Pollution-TSAP' (PPC, 2008) which require a massive reduction in emissions by 2020, as well as the commitments undertaken by the country regarding climate change.

Kyoto Protocol and the targets that must be met

Under the Kyoto Protocol, as noted above (see introduction), developed countries are required to take measures to reduce their greenhouse gas emissions. This is in line with the consensus that industrialized countries are primarily

Table 10.6 The PPC's targets for the percentage reduction of emissions produced for the production of 1 kWh

Emissions	*2006*	*2015*	*Percent change*
Kg CO_2	1.2	0.9	–25
g SO_2	8.3	0.7	–91
g NO_x	1.7	1.1	–39
g PM	0.7	0.3	–56

Source: PPC (2008).

Table 10.7 Relative changes in the reduction of CO_2 emission relating to the country's sustainability (forecasts for 2050)

Industry	*Emissions: million tons of CO_2*			*Emissions in 2050 relative to 1990 (%)*
	1990	*2005*	*2050*	
Supply of electrical energy	43	58	3	–93
Total	109	139	36	–67

Source: WWF Hellas (2008).

responsible for the greenhouse effect and must therefore take the first steps to limit the various gases.

The European Union undertook the commitment in Kyoto to reduce its greenhouse gas emissions by 8 per cent in 2010 compared with 1990 levels. In the framework of allocating responsibility among the EU member states, Greece successfully pressed to be allowed to increase its emissions by 25 per cent by 2012 (once again, compared with 1990 levels). According to European Environment Agency Technical report No. 4/2009, Greece shows a 24.9 per cent increase in greenhouse gases in 2007 compared with 1990 emission levels.

As far as Greece is concerned, the target set under the terms of the Kyoto Protocol is for a 25 per cent increase in emissions of the six greenhouse gases during the period 2008–2012 compared with 1990 levels (the base year for HFC, PFC and SF_6, is 1995). Although this target allows an increase – in contrast with the majority of member states which are required to reduce emissions (Table 10.8) – it is still difficult to attain given that the upward trend of emissions is leading to an even greater increase.

Greece is bound by the target set in Directive 2001/77/EC[3] to achieve a 20.1 per cent rate of participation of RES in electricity generation by 2010. At the present time, Greece has fallen far behind this European target (8.5 per cent).

It emerges from the above that it will be difficult for Greece to achieve the targets set by the Kyoto Protocol. Despite this, certain assumptions are made in this section aimed at attaining the Kyoto objectives. Specifically, the 8 per cent reduction in CO_2 for the year 2005 will eventually lead to the need to replace lignite with photovoltaic systems for the production of electricity.

The reduction in lignite and its replacement with photovoltaic systems

In order to achieve the emissions target for 2012, it is clear that a reduction of around 8 per cent must be achieved, especially in CO_2 which is the main factor responsible for the greenhouse effect. The 2005 NAMEA table for Greece will be used to calculate the size of the reduction in the lignite used in electricity production. The lignite produced by PPC's mines in 2005 totalled 67.2 million tons (Mtn) (Figure 10.2).

Table 10.8 Greenhouse gas emissions for the period 2008–2012 relative to base year 1990 (%); commitments undertaken by the EU member states to reduce emissions

Austria	–13
Belgium	–7.5
Denmark	–21
Finland	0
France	0
Germany	–21
Greece	25
Ireland	13
Italy	–6.5
Luxembourg	–28
Netherlands	–6
Portugal	27
Spain	15
Sweden	4
United Kingdom	–12.5
Kyoto target for EU15	–8
Kyoto target for EU10	–22.6
Kyoto target for EU25	–10.8

Source: EUROPA – Press Releases (2002).

Lignite deposits (Kavouridis *et al.*, 2005) in Greece are expected to last for the next 40 years, since there are 6.7 billion tons (Gtn), of which only 3.3 billion tons (Gtn) can be used for the production of electricity. The annual production of lignite after the Kyoto Protocol rose by 20 per cent to reach an average of 70 million tons annually. The combustion of lignite in 2005 resulted in the production of a total of 43,040 kton of CO_2.

Table 10.9 shows the installed capacity and corresponding production of electricity for 2005. It can be seen that plants producing electricity from the

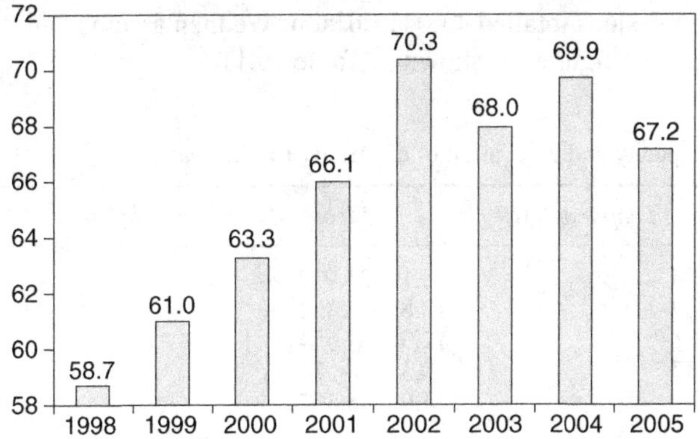

Figure 10.2 Annual production of lignite 1998–2005 (million tons) (source: Kavouridis *et al.*, 2006).

combustion of lignite account for the highest percentage of installed capacity (43.1 per cent), around 5,288 MW, which produces 31,977 GWh of electricity. These are followed by hydroelectric plants which produce 5,367 GWh of electricity, oil-powered plants with 2,309 MW and those operating on natural gas with 1,581 MW of installed capacity, accounting for 15.7 per cent and 8.2 per cent of total electricity production respectively. Lastly, it should be noted that the total installed capacity of RES in 2005 was just 27 MW which produces 56 GWh of electricity and accounts for 0.1 per cent of total electricity generation.

Table 10.10, which is an excerpt of the published NAMEA 2005 table for the Greek economy, shows that 50,316.32 kton of CO_2 resulted from the production of electrical energy in 2005.

Table 10.A.1 (see Appendix) shows that the highest percentage of total CO_2 emissions comes from the generation of electrical energy.

Of these (Kanellopoulos *et al.*, 2007), 43,041 kton of CO_2 are produced by the generation of electricity using lignite as its energy source.

According to the targets that must be achieved (Kyoto target for EU15), we assume that Greece must reduce its CO_2 emissions by 8 per cent. This calculation will be made using the NAMEA table. It should be noted that the intention is for the reduction in CO_2 emissions to be achieved from a decrease in the amount of lignite used to produce electricity by replacing it with photovoltaic systems.

Table 10.11 specifies the amount of lignite that would have to be replaced by photovoltaic systems in 2005.

According to Table 10.11, in 2004, the installed capacity was 5,528.00 MW from which a total of 32,552 GWh of electrical energy was produced from the combustion of approximately 70 million tons of lignite. The CO_2 emissions from the combustion of lignite used to generate electricity totalled 43,814.95 kton.

For the year 2005, on the basis of data published by the PPC, the installed capacity was 5,430.35 MW, from which a total of 31,977.00 GWh of electrical energy was produced from the combustion of 67.3 million tons of lignite. The corresponding CO_2 emissions totalled 43,041.00 kton. We then assumed an 8 per cent reduction in the 2005 figures, as shown in Table 10.11.

Table 10.9 Installed capacity and generation of electricity for the year 2005

Energy sources	*Installed capacity (MW)*	*%*	*Electricity generation (GWh)*	*%*
Lignite	5,288	43.1	31,977	60.4
Oil	2,309	18.8	7,862	14.9
Natural gas	1,581	12.9	7,654	14.5
RES	37	0.3	74	0.1
Hydroelectric	3,061	24.9	5,367	10.1
Total:	12,276	100.0	52,934	100.0

Source: PPC (2006).

Table 10.10 Emissions from the production of electricity, excerpt from NAMEA 2005

Emissions industry	*CO_2 (kton)*	*CH_4 (kton)*	*N_2O (kton)*	*NO_x (kton)*	*CO (kton)*
40–41 electrical energy	50,316.32	0.36573	6.707133	85.19737	48.71095

Emissions industry	*NMVOC (kton)*	*SO_2 (kton)*	*TSP*	*PM_{10}*	*$PM_{2.5}$*	*NH_3*
40–41 electrical energy	13.18597	379.4537	19.97	16.94	10.97	0.09

Source: Economidis *et al.* (2008).

Table 10.11 Installed capacity, electricity generation and emissions per specific quantity of lignite

Year	*CO_2 emissions (kton)*	*Installed capacity (MW)*	*Net electricity generation (GWh)*	*Amount of lignite (million tons)*
2004	43,814.95	5,528.00	32,552.00	69.9
2005	43,041.00	5,430.35	31,977.00	67.3
2005 after the reduction	39,597.72	4,995.92	29,418.83	61.92
Size of reduction	3,443.28 (8%)	434.43	2,558.17	5.38

Source: Kanellopoulos *et al.* (2007).

There must therefore be an 8 per cent reduction in emissions compared with 2005 (i.e. they must be decreased to 39,597.72 kton of CO_2). In order to achieve this 8 per cent reduction in emissions, there must be a reduction of 2,558.17 GWh in net electricity generation compared with initial 2005 levels. In view of the above, the installed capacity must be reduced to 4,995.92 MW and the total amount of lignite used will decrease significantly to 61.92 million tons compared with 2005 when the corresponding amount was 67.3 million tons.

However, demand for electricity is assumed to rise or at least remain the same as 2005 levels. Consequently, this reduction in net electricity generation, which will result from a decrease in the amount of lignite used, will have to be made up for by electricity generated from the installation of photovoltaic systems. According to Table 10.11, the amount of electrical energy which must be generated by photovoltaic system technology has been calculated as being equal to the size of the reduction in electricity generation from the combustion of lignite which is 2,558.17 GWh. According to our calculations, this amount of energy requires the installation of approximately 1,967.8 GW of photovoltaic technology.

Results based on the 2005 NAMEA table

As already mentioned, Table 10.A.1 in Appendix shows an excerpt from the NAMEA Table of Greek Economy for Industry 40–41 in 2005. In the electricity generation industry, the main factor responsible for air pollution is CO_2. According to the 2005 NAMEA table, the amount of CO_2 for 2005 was 50,316.32 kton.

We assume that in order for Greece to achieve the target set by the Kyoto Protocol, it must reduce its CO_2 emissions by 8 per cent. Consequently, CO_2 emissions for the electrical energy industry after the 8 per cent reduction are 46,291.01 kton (Table 10.12).

The change in CO_2 emissions measured in physical units results in a change in the direct coefficients of intensity of air emissions (a_{kj}) as well as in the physical amount of emissions of the GWP group that is symbolized by (E_{kj}) and relates to climate change and the greenhouse effect.

According to the results for the calculation of the direct coefficient (a_{kj}), electricity generation is the industry that is primarily responsible for global warming (Tables 10.13 and 10.14). The direct coefficient of intensity for the GWP group of the electrical energy industry is 11.4995 whereas the emissions caused by the generation of electricity total 52,403.21 kton.

According to the assumption of an 8 per cent reduction in CO_2 emissions, we get new values for the direct coefficients and the emissions per category of air pollution. That is, the direct coefficient of intensity for the GWP group of the electrical energy industry becomes 10.6162 whereas the emissions caused by the generation of electricity total 48,377.906 kton.

Table 10.12 Pollution emissions for industry 40–41, after the 8% reduction in the year 2005

	CO_2	CH_4	N_2O	NO_x kT	CO kT	NMVOC kT	SO_2 kT	TSP	PM_{10}	$PM_{2.5}$	NH_3
40–41	46,291.01	0.37	6.71	85.20	48.71	13.19	379.45	19.97	16.94	10.97	0.09
Total	88,129.12	391.79	40.94	357.35	614.37	356.38	543.77	67.67	47.58	32.48	47.13

Table 10.13 Quantity of emissions by category of environmental pressure variables and direct coefficients of intensity of air emissions for industry 40–41

	Ekj			*akj*		
	GWP	*ACID*	*TOFP*	*GWP*	*ACID*	*TOFP*
40–41	52,403.21	439.26	122.49	11.50	0.096	0.027

Table 10.14 Quantity of emissions by category of environmental pressure variables and direct coefficients of intensity of air emissions after the 8% reduction

	Ekj			*akj*		
	GWP	*ACID*	*TOFP*	*GWP*	*ACID*	*TOFP*
40–41	48,377.91	439.26	122.49	10.62	0.096	0.027

With regard to the air pollution categories ACID and TOFP, the data remained unchanged since the change stemming from the reduction in CO_2 does not affect them.

In Tables 10.15 and 10.16, the total environmental impact is calculated according to relationship eight as well as the elasticity of emission intensity for production e_j^p and final demand e_j^c.

According to Table 10.14, it can be seen that a change in CO_2 emissions from electricity generation can cause changes in the direct coefficients of intensity and, by extension, in the total coefficients (Table 10.16).

Following an 8 per cent change in CO_2 emissions, the electrical energy industry has a smaller effect on all the other sectors in the economy. The total environmental impact (direct and indirect) for the electrical energy industry on the industry itself was reduced from 12.372 (Table 10.15) to 11.4219 (Table 10.16).

At the same time, reductions can be seen in the data on the elasticity of emission intensity for production e_j^p and final demand e_j^c. It can be seen that e_j^p for the total electrical energy industry changes from 12.4757 to 11.5254 and e_j^c from 17.1905 to 15.8701.

Lastly, the direct and indirect emission intensity per unit of final demand categories is computed. Where ε^d is the industry's extent of each emission factor intensity on each final demand component of the Greek economy for the year 2005.

The results we obtain from Tables 10.17 and 10.18 show that the direct and indirect emission intensity per unit of total final demand for the electrical energy

Table 10.15 Total coefficients (direct and indirect) of air emissions. The elasticity of emission intensity with respect to final consumption e_j^c and elasticity of emissions intensity with respect to production e_k^p

	01 & 02	*5*	*10, 11 & 12/13–14*	*15–16*	*17–19*	*20A*	*21–22*	*23*
40–41	0.2656	0.0310	0.1288	0.1966	0.3415	0.2656	0.2231	0.1460
e_k^p	1.6502	0.3420	0.3950	0.8019	0.5094	0.3938	0.3498	0.6986

	24–25	*26*	*27*	*28*	*29–36*	*37*	*40–41*
40–41	0.1989	0.5163	0.7302	0.2840	0.0667	0.1633	12.3723
e_k^p	0.4632	3.6759	1.4029	0.7318	0.2324	0.4496	12.4758

	45	*50–52*	*55*	*60–64*	*65–67*	*70–74*
40–41	0.1575	0.0989	0.2438	0.1403	0.1610	0.0322
e_k^p	0.5694	0.1678	0.3844	0.3995	0.2044	0.0651

	75&90	*80–84*	*85*	*91*	*92, 93, 95 & 99*	e_j^c
40–41	0.1449	0.0759	0.1081	0.0977	0	17.19
e_k^p	0.3969	0.1157	0.1656	0.1421	0.0783	27.26

Table 10.16 Total coefficients (direct and indirect) of air emissions. The elasticity of emission intensity with respect to final consumption e_j^c and elasticity of emissions intensity with respect to production e_k^p, after the 8% reduction

	01 & 02	*5*	*10, 11 & 12/13–14*	*15–16*	*17–19*	*20A*	*21–22*	*23*
40–41	0.2452	0.0287	0.1189	0.1815	0.3153	0.2452	0.2060	0.1348
e_k^p	1.6298	0.3396	0.3851	0.7868	0.4832	0.3734	0.3327	0.6874

	24–25	*26*	*27*	*28*	*29–36*	*37*	*40–41*
40–41	0.1837	0.4767	0.6741	0.2622	0.0616	0.1508	11.4219
e_k^p	0.4479	3.6362	1.3468	0.7100	0.2273	0.4370	11.5254

	45	*50–52*	*55*	*60–64*	*65–67*	*70–74*
40–41	0.1454	0.0913	0.2251	0.1295	0.1486	0.0297
e_k^p	0.5572	0.1602	0.3657	0.3887	0.1921	0.0626

	75 & 90	*80–84*	*85*	*91*	*92, 93, 95 & 99*	e_j^c
40–41	0.1338	0.0701	0.0998	0.0902	0.0000	15.8701
e_k^p	0.3858	0.1099	0.1573	0.1346	0.0783	25.9409

Table 10.17 Direct and indirect emissions intensity per unit of final demand categories

	Final consumption of households	*Final consumption of non profit institutions*	*Final consumption of general government*	*Gross fixed capital formation*	*Changes in inventories*	*Exports*	*Final demand*
40–41	0.3817	0.0005	0.0359	0.0479	0.0003	0.0500	0.5163
Sum	0.6228	0.0008	0.0859	0.1627	0.0005	0.1396	1.0124

Table 10.18 Direct and indirect emissions intensity per unit of final demand categories, after the 8% reduction

	Final consumption of households	*Final consumption of non profit institutions*	*Final consumption of general government*	*Gross fixed capital formation*	*Changes in inventories*	*Exports*	*Final demand*
40–41	0.3524	0.0005	0.0331	0.0442	0.0002	0.0462	0.4767
Sum	0.5935	0.0008	0.0832	0.1590	0.0005	0.1358	0.9727

industry initially is 0.5163 whereas with the assumption of an 8 per cent reduction in GWP group emissions, this coefficient fell to 0.4767.

Household final consumption is the most important category of final demand for the electrical energy industry since it contributes the highest percentage. Thus, the per-unit change in household final consumption initially caused 0.3817 direct and indirect emissions for the electrical energy industry whereas after the 8 per cent reduction in CO_2 emissions, this coefficient fell to 0.3524.

Conclusions

In this chapter we have made calculations that relate to Greece. If the country is to avoid the imposition of sanctions due to its failure to achieve the targets set in the Kyoto Protocol, it must reduce its CO_2 emissions, which are responsible for the greenhouse effect, by 8 per cent.

On the basis of data taken from the PPC's operational programme, we have made certain calculations to reduce CO_2 emissions to 39,597.72 kton. The combustion of lignite for electricity generation (8 per cent) must be followed by a reduction in net electricity generation of 2,558.17 GWh compared with 2005 levels. Installed capacity must be reduced to 4,995.92 MW and the total amount of lignite used must be significantly reduced to 61.92 million tons compared with 2005 when the corresponding amount was 67.3 million tons. Lastly, the demand for electricity is assumed to rise or at least remain the same as 2005 levels. Consequently, this reduction in net electricity generation, which will result from a decrease in the amount of lignite used, will have to be made up for by electricity generated from the installation of photovoltaic systems. The amount of electrical energy which must be generated by photovoltaic systems technology has been calculated as being equal to the size of the reduction in electricity generation from the combustion of lignite which is 2,558.17 GWh. According to our calculations, this amount of energy requires the installation of approximately 1,967.8 GW of photovoltaic technology.

According to the 2005 NAMEA table, the main factor responsible for air pollution is CO_2 and 50,316.32 kton are produced in the electrical energy industry. CO_2 emissions in the electrical energy industry after the 8 per cent reduction are 46,291.01 kton. This change in turn results in a change in the direct coefficients of intensity of the GWP group which had initially been estimated at 11.4995, whereas after our assumption, this figure fell to 10.6162. The change in the direct coefficients of air emission intensity results in a change in direct and indirect impact data and, by extension, in the elasticity of emission intensity related to production and final demand.

The total environmental impact (direct and indirect) for the electrical energy industry was therefore reduced from 12,3722 to 11,4219, whereas the elasticity of emission intensity relative to production for the electrical energy industry fell from 12,4758 to 11,5254 and the elasticity of emission intensity for final demand changed from 17,1906 to 15,8701. The magnitude of the coefficients means that the results are significant.

Lastly, the direct and indirect emission intensity per unit of final demand categories is computed. The change in the total final demand per unit for the electrical energy industry initially caused 0.5163 direct and indirect emissions whereas after our assumption (of an 8 per cent reduction in CO_2 emissions), this coefficient fell to 0.4767.

Household final consumption is the most important category of final demand for the electrical energy industry since it contributes the highest percentage. Thus, the per-unit change in household final consumption initially caused 0.3817 direct and indirect emissions for the electrical energy industry whereas after the 8 per cent reduction in CO_2 emissions, this coefficient fell to 0.3524.

Appendix

Table 10.A.1 Excerpt from NAMEA table of Greek economy for industry 40–41 in the year 2005 (emissions)

CO_2 kT	CH_4 kT	N_2O kT	NO_x kT	CO kT	
50,316.32	0.37	6.71	85.20	48.71	
NMVOC kT	SO_2 kT	*TSP*	PM_{10}	$PM_{2.5}$	NH_3
13.19	379.45	19.97	16.94	10.97	0.09

Source: Economidis *et al.* (2008).

Notes

1 Law 3468/2006 on the 'Generation of Electricity Using Renewable Energy Sources and High-Efficiency Cogeneration of Electricity and Heat and Other Provisions'.
2 Speech on 'Investments: Renewable Energy Sources' by the Ministry General Secretary at a meeting organized by the Chamber of Light Industry, Piraeus, 2007.
3 Directive 2001/77/EC deals with the production of electricity from RES. The purpose of the directive is to promote an increase in the contribution of RES to electricity production in the internal market for electricity and to create a basis for a future Community framework.

References

de Haan, M. and Kee, P. (2004) 'Accounting for sustainable development: the NAMEA-based approach', in Statistics Netherlands (ed.), *Measuring Sustainable Development Statistics*, Paris: OECD, pp. 183–199.

de Haan, M. and Keuning, S.J. (1996) 'Taking the environment into account: the NAMEA approach', *Review of Income and Wealth*, 42: 131–148.

de Haan, M., Keuning, S.J and Bosch, P.R. (1994) 'Integrating indicators in a National Accounting Matrix including Environmental Accounts (NAMEA)', National Accounts Occasional Papers NA-60, Voorburg/Heerlen: Statistics Netherlands.

Economidis, C., Keramidas, D., Demertzi, A., Stromplos, N. and Vlachogianni, D. (2008) 'The compilation of a Greek environmental input–output table for 2005', paper presented at the International Input–Output meeting on managing the environment, International Input–output Association, Seville. Online: www.iioa.org.

Environmental, Education, Sustainability (2007) 'Photovoltaics: a practical guide'. Online: www.env-edu.gr/Documents [in Greek].

EUROPA – Press Releases (2002) 'The Kyoto protocol and climate change: background information'. Online: http://europa.eu/rapid/climat/PressReleases Action.

European Environment Agency (2008) *EN09 Emissions (CO_2, SO_2 and NOx) from Public Electricity and Heat Production: Explanatory Indicators*, Copenhagen: EEA.

European Environment Agency (2009) 'Annual European Community greenhouse gas inventory 1990–2007 and inventory report 2009', Technical Report No. 4/2009, p. 9, Copenhagen: EEA.

European Environment Agency (2010) 'Annual European Community greenhouse gas inventory 1990–2008 and inventory report 2010', Technical Report No. 6/2010, p. 10, Copenhagen: EEA.

Eurostat (2001) 'NAMEAs for air emissions 2001 edition' (Detailed Tables Theme 2 Economy and Finance), Luxembourg: Office for Official Publications of the European Communities, p. 11. Online: http://epp.eurostat.ec.europa.eu.

Eurostat (2007) 'NAMEA for air emissions', Compilation Guide, Preliminary Draft Version to be discussed and further developed by the ESTAT NAMEA TF, European Community, Luxembourg: Eurostat.

Greenpeace (2006) 'The end of lignite'. Online: www.greenpeace.org [in Greek].

International Energy Agency (2006) 'Compared assessment of selected environmental indicators of photovoltaic electricity in OECD cities', Report IEA-PVPS T 10-01, May.

IPCC (2007) *Climate Change 2007 Synthesis Report*, Valencia, Spain.

Kanellopoulos, D., Kromba, D. and Perdikaris, N. (2007) 'Tackling the phenomenon of climate change in electricity generation', *IENE – PPC*. Online: www.dei.gr/Documents/IENE-DEH pdf [in Greek].

Katsouyanni, K., Touloumi, G. and Samoli, E. (2001) 'Confounding and effect modification in the short-term effects of ambient particles on total mortality: results from 29 European cities within the APHEA2 project', *Epidemiology*, 12: 521–531.

Kavouridis, K. (2008) 'Lignite industry in Greece within a world context: mining, energy supply and environment', *Energy Policy*, 36: 1257–1272.

Kavouridis, K., Galitis, N. and Karampakakis, J. (2006) 'Health and safety management in the open pit lignite mines of Public Power Corporation', *Mining and Metallurgical Chronics*, May [in Greek].

Kavouridis, K., Haloulos, K., Leontidis, M. and Roubos, C. (2005) 'Lignite and natural gas in the country's electricity production', Athens: Technical Chamber of Greece [in Greek].

Leontief, W. (1970) 'Environmental repercussions and the economic structure: an input–output approach', *Review of Economics and Statistics*, 52: 262–271.

Leontief, W. (1986) *Input–Output Economics*, New York: Oxford University Press.

Miller, E.R. and Blair, P.D. (2009) *Input–Output Analysis: Foundation and Extensions*, 2nd rev. edn, Cambridge: Cambridge University Press.

Ministry of the Environment, Physical Planning and Public Works (2005) 'Stocktaking of gases, pollutants, liquid and solid waste from industry and emission from central heating: processing of stocktaking results – calculation of emission coefficients'. Online: www.minenv.gr [in Greek].

Monthly Bulletin, Hellenic Transmission System Operator (2008) November. Online: www.desmie. gr [in Greek].

Mylonas, N., Economakou, M., Frankopoulos, N., Krasadakis, A., Molfetas, K., Stromplos, N. and Vlachos, P. (2000) *Natural resource and environmental input–output tables for Greece, 1988–98*, Athens: Eurostat and Institute of Computer and Communications Systems (ICCS) of National Technical University of Athens, pp. 58–63 [in Greek].

National Observatory of Athens (2002) *Climate Forecasts 2071–2100*, Athens: National Observatory of Athens [in Greek].

National Observatory of Athens and Ministry of the Environment, Physical Planning and Public Works (2001–2002) 'Examining support policies to advance the ministry's policy measures aimed at reducing CO_2 emissions'. Online: www.energycon.org/OEE-Overview pdf [in Greek].

PPC (2008) 'The role of coal in PPC's production strategy', Athens: PPC. Online: www.dei.com.gr/Documents/dei-anthrakas.pdf [in Greek].

PPC (2006) *Report Year 2005*, Athens: PPC, pp. 46, 47 [in Greek].

UNFCCC (2008) *Kyoto Protocol Reference Manual on Accounting of Emissions and assigned Amounts*, Bonn: UNFCCC.

WWF (2005) 'Climate change impacts in the Mediterranean resulting from a 2 degrees C global temperature rise', Gland: WWF.

WWF Hellas (2008) *Solutions for Climate Change: Vision of Sustainability for Greece in 2050*, Scientific Report of WWF Hellas [in Greek].

11 Index-based decomposition of SO_2, NO_x, CO and PM emissions stemming from stationary emission sources in the Czech Republic over 1997–2007

Milan Ščasný and Fusako Tsuchimoto Menkyna

Introduction

Decomposition analysis has been widely employed to examine the effect of driving forces on the variables in question for over 30 years. At the very beginning of its application in the field of environmental economics, it was intended to break down changes in aggregated energy intensities into changes in the production mix of industry and changes in sectoral energy intensity. Since then, this technique has become one of the most common methods in the analysis of energy or emission trends. In fact, when Tol and Weyant (2006) searched for the most cited papers published in the *Journal of Energy Economics*, they found that eight of the top ten papers are on decomposition methods and its applications.

A decomposition analysis differs from an econometrical analysis which aims to estimate a reduced form of relationship between a dependent variable and covariates. A decomposition analysis aims to explain the channel through which certain factors affect a variable. In other words, decomposition cannot establish causal linkages between the driving force and the dependent variable, but it can examine the underlying causes of certain factors. How can we interpret the result of decomposition? For instance, Seibel (2003) interprets the effect of certain factors of decomposition as what would have happened if only the factor under consideration had changed while all the others had not. In general, each factor captures how much of the change in the dependent variable that is observed during a certain period can be assigned to changes in each of the factors. Löfgren and Muller (2010) claim that it is the magnitude of the contributions of each factor that determines the effect of changes in the driving force if nothing else has changed. In other words, they consider that the value of a decomposition analysis lies in the indication of what may and may not be important in the future trend in the indicator in question.

Most decomposition studies, and particularly index-based analysis, have been applied in the former EU15 countries and Asia, mainly in China, with some applications in the US and Canada, and selected OECD or International Energy Agency (IEA) countries (Löfgren and Muller, 2010). In fact, very few applications of the decomposition analysis have been performed for African countries, which might be due to the scarce availability of data. We also find that there are few applications in Central and Eastern Europe. For instance, Viguier (1999) and Cherp *et al.* (2003) decompose air emissions for several Central and Eastern European countries, whereas Brůha and Ščasný (2006) or Tsuchimoto and Ščasný (2011) decompose changes in emissions year by year for the Czech Republic. Vehmas *et al.* (2008) apply the decomposition analysis to examine changes in several indicators in a few of these countries.

The aim of this study is to contribute to this literature by an application of the decomposition analysis in one country in Central and Eastern Europe. Specifically, our analysis focuses on driving forces of changes in air emission levels during the period of economic and political transformation in the Czech Republic. Before 1990, during the period of centrally planned system, the Czech economy was characterized by inefficiencies in the economy, a large share of heavy industry and high resource and pollution intensities due to a lack of effective environmental regulation. For instance, total emissions of particulate matters or SO_2 were over ten times, or eight times respectively, larger in 1990 (i.e. the first year of the transformation) than ten years later.

Transformation of the system led to an introduction of significant changes in economic structure, institutions and governance. As a result of new governance, a new Air Quality Law was introduced in 1991 which required each large stationary emission source to fulfil quite strict emission limits until 1999. Since 2000, the Czech government has also introduced several more policies to combat air pollution in order to comply with European legislation. Hand in hand with process and technological changes in the Czech economy, all of these structural and institutional changes resulted in a significant reduction in air emissions. The aim of this chapter is to shed light on what the key underlying cause was during the transformation and post-transformation period in the Czech economy of reducing air emissions stemming from stationary emission sources.

The specific objective of our study is three-fold. First, we undertake a four-factor index decomposition analysis linked to the Divisia index to break down changes in emission levels of SO_2, NO_x, CO and particulate matters into scale, composition, fuel intensity and emission–fuel coefficient effect. We also examine the value added of undertaking the four-factor decomposition analysis compared with quite standard three-factor analysis. Second, the decomposition analysis may provide a biased result if the economy is broken down into few economic sectors. For this reason, we undertake our decompositions at various levels of disaggregation of the Czech economy in order to examine what the effect on the result might be if various sector resolutions are applied in this decomposition analysis. Third, we focus more on time aggregation. First of all, we aggregate the decomposition result as based on year-by-year changes for three distinct time periods to examine a trend in driving forces of air emission changes during the period when firms were obliged to fulfil strict requirements on emission limits (1997–2000), during the period when government and enterprise efforts were determined by implementation of EU *acquis communitaire* (2000–2004) and lastly, during the period when enterprises had to do their best to become competitive in the EU market (2004–2007). Changes in an indicator in question over a certain period can in principle be decomposed by using a period-wise method in that only values for the first and final year in the period are considered. Alternatively, we can decompose the changes year by year. Our last concern here is to document for changes in particulate matters in the Czech Republic, what differences in decomposition results there might be if we undertake decomposition on a year-by-year basis and if decomposition analysis is based only on the first and final years of the period.

Decomposition in brief

Two decomposition techniques

In brief, a decomposition analysis basically looks at the difference between a (dependent) variable at two different points in time and attributes the changes in that variable to several determinants that are called factors or underlying driving forces.

There are two basic streams of decomposition analysis. S*tructural decomposition analysis* (SDA) describes decomposition studies based on input–output models, whereas *index decomposition analysis* (IDA) relies on index theory. Both of the methods developed side by side but independently of each other from the mid-1970s until Hoekstra and van den Berg (2003) compared both decomposition techniques and established a link between them.

Let us first discuss SDA. It is based on input–output data and can distinguish between a variety of technological and final demand effects which IDA is not able to detect. According to Jungnitz (2008), SDA is also less limited than a standard input–output analysis with fixed technological or production coefficients; in principle, SDA can identify the effect of changing those coefficients on the variable of the analysts' interest. On top of the direct effect from changing production in a particular sector, structural decomposition can also capture indirect effects (i.e. the effect of a direct change in demand in one sector on the inputs of other sectors through the Leontief inverse). The main obstacle for conducting a structural decomposition analysis is the lack of availability of input–output tables for each year in the analysed period, which is not often the case.

IDA, on the other hand, merely considers the direct effects, but requires less data and that is also the reason why this technique has become so popular and widely used in empirical work. This is also why our study relies on using the IDA rather than the SDA approach. Furthermore, we discuss the indexes and other methodological considerations relevant to the IDA approach in more detail.

Index decomposition analysis and choice of index

The starting point of any decomposition analysis is to set an equation through which an analyst defines the relationship between some dependent variables and several 'underlying causes'. Such underlying causes are called factors. A specific feature of such an equation is that the product of all the factors has to equal the variable, the change of which is actually analysed in the decomposition. Often, the chosen factors are simply the ratios where the denominator of one factor equals the numerator of the next one, as shown here for a two-factor decomposition analysis:

$$X = \frac{X}{Y} \times Y. \tag{1}$$

The choice of factors should be determined by the analyst's conceptual model. In the decomposition of an environmental burden, and emissions in particular, analysts often decompose change in the environmentally related variable into three factors. The first, *scale* (or sometimes *production*) *factor*, measures the effect of the growth of the overall economy (e.g. doubling the economic performance will double the environmental burden). The second, *composition* (or *structural*) *factor*, relates to changes in the structure of the economy (usually measured via the share of sectoral value added or production in overall value added, or production respectively) holding the scale of economy and technologies constant. Lastly, the *intensity* (or *technique*) *factor* describes the input intensity of a sector's production, e.g. the emission intensity to produce a unit of value added. Some studies go further and conduct a four-factor decomposition analysis by decomposing the intensity effect further into two factors which are energy intensity and emission coefficients, or more factors.[1] For example, using a unique dataset for Czech emission sources, Tsuchimoto and Ščasný (2011) decompose the intensity effect into energy intensity, fuel mix and emission coefficients, with five factors in their decomposition study.

There are two main streams of decomposition analysis based on the choice of index.[2] First, there is the Laspeyres index that uses base year weight and the Paasche index with target year weights belonging to one stream of decomposition indexes, and both can be applied in structural and index decomposition analyses.[3] In both cases, decomposition generates a residual term which captures a 'mixed effect' (Seibel, 2003) caused by an interaction among driving forces. It basically implies that the aggregate is not perfectly decomposed into isolated effects that are also considered to be the main disadvantage of this approach.[4]

In general, there are three options for dealing with the residual: we can either specify decomposition in a way that the residuals are avoided (see the Divisia index), we can only consider the isolated effects (Seibel, 2003) or we can accept the residuals. In the third case where the residuals are accepted, the next question is how to address them.

Seibel (2003) considers three main possibilities of addressing the residuals. The first option is to ignore the residual term altogether. We can only ignore the residual term if it is quite small compared to the overall change in the dependent variable. The second option is to consider the residual term in the analysis. Lastly, the residual term can be distributed among the different isolated effects by *using some allocation rule*. For instance, the so-called *Sun/Shapley method*

(Sun, 1998, 1999) *sets the allocation weight* for a two-factor decomposition equal to one-half, which basically means that the residuals are allocated equally among the factors. In the modified version of the Sun/Shapley method, as used in the DE-COIN project (Vehmas *et al.*, 2008), the weight is allocated according to the contribution of each factor towards the total amount of the change in the variable decomposed. In both cases, the residuals disappear but the problem still remains because it is basically the analyst who allocates the residuals to each factor. The result thus simply depends on the analyst's arbitrary decision on how residuals are allocated to the factors.

The Divisia index-based decomposition then belongs to the second stream of decomposition analysis. Its main advantage is that Divisia index-based approaches do not generate residuals, or, practically speaking, it only creates negligible residuals. However, as argued by Jungnitz (2008), this method can only be applied to the index decomposition approach and not to SDA.

Selection of the index decomposition method between the Laspeyres/Paasche index and the Divisia index basically implies whether one prefers to specify the decomposition in a way that the residuals are possible or avoided.

In both IDA approaches as linked to the Laspeyres/Paasche or to the Divisia method, a change in the dependent variable as well as in the driving forces (decomposition factors) can be expressed either in absolute or in relative terms. Decomposition itself can then have either a multiplicative or additive form, i.e. changes in a single driving force can be linked multiplicatively or additively.[5] In the former form, the change in the dependent variable is decomposed into the product of the change in each factor whereas in the latter, the change is decomposed in the sum of changes attributable to each of the factors. The additive form of decomposition has an advantage in that interpretation of its results is more straightforward than for a multiplicative form of decomposition. In the additive form of decomposition, the magnitude of change in the dependent variable simply equals the sum of the magnitude of change in the factors into which the variable is decomposed.[6]

Many, for instance Jungnitz (2008), argue that each approach to undertaking IDA – based on the choice of index, on a multiplicative or additive form, or on relative or absolute changes – is consistent with statistical theory and therefore there can be no unique recommendation for a decomposition approach (e.g. Diekmann *et al.*, 1999).

However, Ang (2004) argues that the choice of IDA method should mainly be based on factor and time reversibility and the most important test is factor reversibility that requires full decomposition with no residuals remaining. Liu and Ang (2007) also provide a coherent framework for an evaluation of index decomposition and a recommendation on methods that meet four criteria for the method selection.[7]

In this regard, the conventional Laspeyres or Paasche-based index decomposition is not recommended because it can create huge residuals. However, although the modified versions of the Laspeyres index (Sun/Shapley method) overcome these shortcomings and possess the full decomposition property, the problem of residuals still remains due to arbitrariness in allocating the residual term among the factors. Furthermore, a linkage between multiplicative and additive property is not clear in the decomposition linked to the Laspeyres index.

The logarithmic mean Divisia index (LMDI) approach is currently seen as the most favourable and the best decomposition method (Ang, 2004; Ang and Liu, 2001; Liu and Ang, 2007; Muller, 2007; Löfgren and Muller, 2010). Therefore, our analysis presented here is also based on LMDI.

Further considerations in IDA: sector and time resolution

There are conclusive methodological recommendations to be emphasized: decomposition results can be sensitive to the level of aggregation (Ang, 1995, 2006) and the decomposition analysis should therefore be undertaken on as detailed a level as possible (e.g. Jungnitz, 2008; Löfgren and Muller, 2010).

Let us first consider the disaggregation level. Generally, a decomposition analysis can be undertaken at any level of the economy including region, country, sectoral, intra-sectoral or even enterprise level. Most of the applications conduct decomposition at national economy level when decomposition is based either on overall figures or each driving force is broken down by sectors and only subsequently the sectoral effects of each factor are aggregated to the overall figure. As shown by Rørmose and Olsen (2003) for the latter case, when decomposing air emissions in Denmark 1980–2001 by employing the SDA technique, the more disaggregated the driving forces are, the larger the effects identified by the decomposition analysis. Similarly, Seibel (2003) applies SDA to the decomposition of CO_2 emissions for Germany and finds that the decomposition results differ when the SDA is based on differently disaggregated data; broken down into 70, 58 or 12 branches. He finds that decreasing the level of sector disaggregation might lead to a biased statement of the intensity-related and the structure-related effects.[8] For example, if the economic output of a less polluting-intensive sector is growing and the polluting sector is declining, then part of the observed emission reduction is caused by this structural change and it shall thus be assigned to the composition effect. As a result, if those two sectors are recorded separately, the intensity effect will be lowered and thus be smaller compared with a situation in which those two sectors are kept in one sector. In the opposite situation, while still assuming an overall emissions reduction, if the economic output of the pollution-intensive sector now increases, then the intensity effect must be bigger after a finer sector breakdown than when there is no such breakdown to offset the negative structural effect.

As described by Löfgren and Muller (2010), when decomposing the change in the dependent variable into the change in factors over a certain period, the result of decomposition is given by an integral of factor functions summed across sectors. The data required to calculate the exact integrals is not, however, available. In line with their argument, we therefore need to rely on the information available for the boundaries given by the initial and target points in time.

Due to the fact that the exact shape of the function between these two points is unknown, each term in the decomposition needs to be somehow approximated. How the period of interest is considered in the analysis will then affect the decomposition result.[9]

In principle, there are two distinct strategies for considering a time period. One can apply a period-wise approach; this basically means that the decomposition study considers only the start and end year of the analysed period. Alternatively, the period of interest can be cut into several consequent boundaries in order to conduct the decomposition separately for each of the sub-periods, most often on a year-by-year basis. The period-wise approach neglects a trajectory path both for each driver and for the dependent variable in the analysis. It basically assumes that both the base year and the final year are representative averages of the trend between the two years (Nanduri, 1996). Due to this fact alone, Löfgren and Muller (2010) recommend that decomposition analysis should not only be based on the first and final years in a long time period. Despite the known shortcomings of the period-wise approach, as noted by Jungnitz (2008), this procedure is followed by several statistical offices.

The aim of this study is to document what might be the effect of various levels of sector disaggregation and basing decomposition on the first and final year of a reasonably long period. In both cases, we apply the IDA approach to decompose air emissions in the Czech Republic during the period 1997–2007.

Our approach

Logarithmic mean Divisia index

Given the reasons mentioned above, we use LMDI decomposition analysis in our study. This method satisfies the property of perfect decomposition, i.e. no residuals are generated and there is also a clear linkage between the additive and the multiplicative decomposition.

In the IDA linked to the Divisia index, the total change in the dependent variable – the overall level of pollution, for example – can be decomposed multiplicatively or additively. Multiplicative LMDI decomposition of overall change in pollution into n factors is given by this formula:

$$E_{total} = \frac{E_T}{E_o} = D_{x1} \cdot D_{x2} \cdot D_{x3} \cdot \ldots \cdot D_{xn} \cdot D_{resid} \tag{2a}$$

where E_{total} is total magnitude of the change in the emission level given by the ratio of the level in the final and initial period, D_x denotes the contribution of factor xk to the change in variable E where $k = 1, \ldots, n$ and n denotes the number of factors the change is decomposed into, and D_{resid} is the contribution of the residual term.

The effect of *kth* factor in multiplicative decomposition can be expressed by:

$$D_{xk} = \exp\left(\sum_i \frac{L(E_{i,T}, E_{i,o})}{L(E_T, E_o)} \ln\left(\frac{x_{k,i,T}}{x_{k,i,o}}\right)\right) \tag{2b}$$

where i denotes sectors and $L(a;\ b)$ is the logarithmic average of the two numbers, a and b, i.e.

$$L(a,b) = \frac{a-b}{\log a - \log b}, \text{ if } a \neq b, \text{ otherwise } L(a, b) = a. \tag{2c}$$

In the additive LMDI decomposition, the change in emission level can be decomposed in the following way (Ang, 2004):

$$\widehat{E_{total}} = E_T - E_o = \widehat{D_{x1}} + \widehat{D_{x2}} + \widehat{D_{x3}} + \ldots + \widehat{D_{xn}} + \widehat{D_{resid.}} \tag{3a}$$

where $\widehat{D_{xk}}$ denotes the contribution of *kth* factor to the change in emission level and equals

$$\widehat{D_{xk}} = \sum_i L(E_{i,T}, E_{i,o}) \ln\left(\frac{x_{k,i,T}}{x_{k,i,o}}\right), \tag{3b}$$

The multiplicative and additive decomposition share the following property for each k:[10]

$$\frac{\widehat{E_{total}}}{\ln E_{total}} = \frac{\widehat{D_{xk}}}{\ln D_{xk}}. \tag{4}$$

However, as the magnitude of the total change becomes larger, the interpretation of the results becomes less intuitive with multiplicative LMDI. To interpret our result easily, we use the additive LMDI in our decomposition analysis.

Because of the logarithmic function in the equation, any analysis is sensitive to zero or negative values in the dataset. In order to deal with this problem, following Ang (2004), we replace a negative or zero value with very small values, i.e. in the range of 10^{-10} to 10^{-20}. Note that this replacement is only valid when the following conditions are satisfied (Wood and Lenzen, 2006): (a) zero values do not outnumber non-zero values, (b) the dataset does not contain many small values, and (b) there are no large changes in the dataset. Looking at the summary statistics of our data set, we fortunately do not have the problem described, and we therefore replace a negative or zero value with 10^{-10}–10^{-20} as Ang suggests.[11]

Three-factor and four-factor decomposition

The change in emission level is decomposed into three (equation 5a) and then into four factors (equation 5b) and following specifically the equations (3a) and 3b):

$$\Delta_t E = \sum_i \left(\Delta_t \frac{E_i}{Y_i} + \Delta_t \frac{Y_i}{Y} \right) + \Delta_t Y \tag{5a}$$

$$\Delta_t E = \sum_i \left(\Delta_t \frac{E_i}{F_i} + \Delta_t \frac{F_i}{Y_i} + \Delta_t \frac{Y_i}{Y} \right) + \Delta_t Y \tag{5b}$$

where $\Delta_t X$ denotes the percentage change in X variable at time t.[12]

The last term in both equations captures the scale effect. It basically shows what change in emission levels could be attributed to changes in overall economic performance. In the three-factor decomposition, there are then two other factors: the composition effect and the emission-intensity effect. The four-factor decomposition then decomposes the latter one further into the fuel-intensity and the emission–fuel coefficient effect (for the detailed description of the variablessee Table 11.1).

Emissions can be, in principle, abated through four distinct channels. Reducing economic production is the first of them. In decomposition analysis, the reduction in the whole economy is directly captured by the *scale effect* whereas the relative reduction of polluting industries is captured in the *composition effect*. The negative values of the composition effect therefore indicate a shift in the economy from dirty sectors towards cleaner ones.[13]

On top of the structural changes in the economy as a whole, emissions can be reduced by reducing fuel consumption. Reductions in fuel intensity due to the use of more fuel efficient technologies (or more efficient energy generation) would directly result in fewer emissions. The effect of technological and/or product change can be captured simply by the *fuel-intensity effect*. The last two abatement options are the installation of end-of-technologies (i.e. filters) and a fuel switch towards using environmentally friendly energy carriers, e.g. replacing brown coal with natural gas. Both of the effects are captured altogether by our last factor, the *emission fuel coefficient*.[14]

Overall, the *emission-intensity factor* in our three-factor analysis captures, without an output reduction, three abatement options altogether, i.e. it captures abatement through end-of-pipe technology, fuel switch and technological and/or product change. This factor thus indicates the effect of the environmental friendliness of production without distinguishing further through which channel the emission was abated.

Focusing on disaggregation level and time period

Furthermore, after undertaking our four-factor decomposition analysis of annual change in air emissions in the Czech Republic, we examined the effect of using a different disaggregation level of sector breakdown and the difference in the result when the data is aggregated over a longer time span.

Specifically, we first decomposed change in air emissions into four factors, although the economy has now been broken down into its sectors following two-character alphabetical codes (31 sub-sections), letters A to Q (17 sections), and

Table 11.1 The variables and factors used in the decomposition analysis

Variables	
$\Delta_t X = \frac{[(X]_t - X_{t-1})}{X_{t-1}}$	Percentage change in variable/factor X over the year, where the operator Δt suggests the fraction of the change in the variable X from time *t-1* to t, in our case between two consequent years
$E = \sum_i E_i$	E_i denotes emissions arising from fuel combustion in stationary emission sources in sector i and E denotes total emissions arising from that emission source during a given year
$F = \sum_i F_i$	E_i denotes total energy consumption used by all stationary emission sources in sector i and F denotes total energy consumption by that emission source during a given year
$Y = \sum_i Y_i$	E_i denotes gross value added in sector i and Y denotes overall gross value added of the whole economy, i.e. GDP for a given year
Factors	
$\Delta_t Y$	The scale effect: it captures the effect of change in performance of whole economy on the emission change
$\Delta_t \frac{Y_i}{Y}$	The composition effect captures the effect of changes in relative sizes of sectors on the change in emission levels
$\Delta_t \frac{E_i}{Y_i}$	The intensity effect captures the effect of changes in emission intensity, i.e. how much pollutants are released to generate a unit of value added across sectors at emission level
$\left(\Delta_t \frac{E_i}{Y_i} = \Delta_t \frac{F_i}{Y_i} + \Delta_t \frac{E_i}{F_i}\right)$	
$\Delta_t \frac{F_i}{Y_i}$	The fuel-intensity effect captures the effect of changes in fuel intensity, i.e. how much GJ is needed to produce one unit of value added, across sectors
$\Delta_t \frac{E_i}{F_i}$	The emission–fuel coefficient effect captures the effect of emission–fuel intensity, i.e. how much pollutants are released per unit of energy input at emission level

lastly nine wider sector categories.[15] The decomposition result is then compared with our base case, where the economy is disaggregated following the sector resolution based on two-digit NACE codes (i.e. 60 divisions).

In our second exercise, we decompose the change in air emissions for three distinct periods, specifically 1997–2000, 2000–2004 and 2004–2007. While the first of them can be characterized by an attempt by firms to fulfil strict requirements to reduce their emissions, the second period can be characterized by adjustments by institutions and trade patterns to be in compliance with standards of the European Community and the last by full membership of the Czech Republic in the Community. We then examine what the effect of each factor would be if we decompose the change in air emissions on a year-by-year basis which is then aggregated for each period in question and if we decompose the changes for the first and final year in each period.

Data description

The source of emissions and energy data is the Air Pollution Emission Source Register (*Register emisí zdrojů znečištění ovzduší*, REZZO). The REZZO database is compiled by the Czech Hydro-Meteorological Institute and covers so-called classical pollutants such as SO_2, NO_x, CxHy (NMVOC), CO, particulates and other trace pollutants as well as consumption of more than 20 types of fuel for all emission sources in the Czech Republic. Fuel consumption by all stationary emission sources is recorded in the REZZO Register for a total of 24 types of energy carrier.

The REZZO Register distinguishes four broad categories of emission sources for which data are stored. Categories R1 and R2 cover stationary emission sources and a distinction is made according to the thermal output of the source; R1 covers very large and large emission sources defined by thermal output larger than 5 MW and R2 includes medium-sized sources with less than 5 MW of thermal output. In the REZZO Register we can further distinguish two specific sources of air emissions for each stationary emission source. The first category covers the emissions stemming from the combustion of fossil fuels and the second covers the emissions as generated by various kinds of technological processes. Category R3 then reports the emissions released by local units including households and area sources while emissions arising from mobile sources are listed under category R4.

For large and very large emission sources (R1), data are gathered and stored for each facility defined by a stack for combustible processes or by specific technological process whereas data for medium-size emission sources (R2) are reported at firm level. Having information on a firm and the sector the firm belongs to allows us to derive the environmental and energy data at firm or sector level. The REZZO database is also the data source used to compile the Czech Air-NAMEA (see Ritschelová *et al.*, 2001).

The change in emission level can also be due to the changes in the composition of emission sources and/or the process that generates them. In our case, the emissions recorded in the register come from three kinds of emission sources such as stationary, area (including households) and mobile. For emissions coming from stationary sources, releases might then be generated either from combustion of fuel or from a certain technological process without using fuel. Admittedly, our composition effect does not include the compositional change of the emission source or the ratio of emissions from fuel combustion.

Specifically, in this study, we decompose air emissions of particulate matters, SO_2, CO and NO_x stemming from fuel combustion and emitted from very large, large and medium-sized stationary emission sources during the period 1997–2007 (see 'R1+R2 fuel combustion' in Figure 11.1). The amount of emissions examined in our study represents the majority of total SO_2 emissions (about 70 to 85 per cent) and a smaller amount of total NO_x (30 to 45 per cent of the total, but 80 to 85 per cent from stationary sources respectively). Due to the dominant contribution to emissions of particulate matters from transport, PM emissions stemming from combustion processes in very large, large and medium-sized emission sources constitute about 10 to 40 per cent of the total, or 36 to 79 per cent of all emissions stemming from stationary sources, respectively.

Sector-level economic data are taken from the consolidated yearly national accounts compiled by the Czech Statistical Office. Economic performance in our analysis is based on gross value added expressed in real 2000 prices by using the GDP deflator provided by the International Monetary Fund.

Overall, our database covers, on average, 80 per cent of total consumption of coal, gas and oil as reported by the Czech Statistical Office. We note at this point that our database does not cover fuel consumption by mobile sources, i.e. propellants nor fuel use by area sources. Apart from propellants, any difference in the aggregate in our database and the aggregate consumption of fuels as reported by the statistical office is therefore explained by the consumption of fuels by technological processes in stationary sources and by area sources. Lastly, although the register distinguishes many kinds of fuels, we use the overall consumption of fuels in our analysis.

During the period 1997–2007, total gross value added in the Czech Republic increased by 44 per cent. During this period, there was only one year when GDP decreased, and that was in 1998 when it decreased by –0.7 per cent. Since then, GDP continuously grew by around 3 per cent p.a. during the period 2000–2004 and by almost 7 per cent p.a. during the period 2005–2007 (see Figure 11.1). During these periods, the share of agriculture (Section A and B), mining (Section C), transport (Section I) and construction (Section F) on total GVA declined, especially over the period 1997–2000. While the reduction in the first two was due to an overall structural change in the Czech economy, the decline in GVA in the transport and construction sectors was mainly due to the delayed effect of economic recession. On the other hand, as shown in Table 11.2, the share of trade (Section H) until 2000 and the share of power (Section E) and public services (Sections L, M, N, O) has increased, especially since 2002. The increase in value added of the trade sector is an indication of the transition from a formerly centrally planned economy to a market system. The

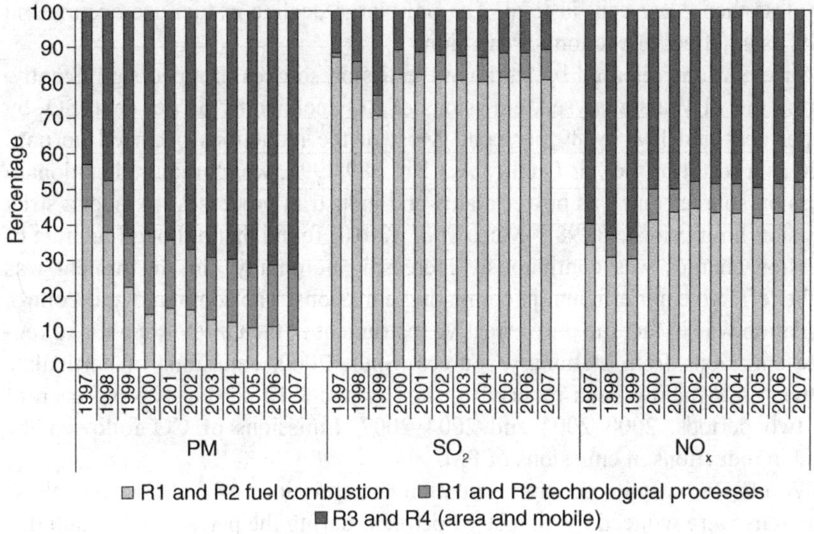

Figure 11.1 Air emissions from various emission sources and processes from 1997 to 2007.

increasing share of power in the total value added was mainly due to the increasing export of electricity as well as the completion of the construction of a second nuclear power plant at Temelín. The increase in the production of public and other services such as health care, education, environmental services and infrastructure and other social, cultural and personal services is due to the increasing wealth and demand of Czech society and because of the activity involved in the fulfilment of EU requirements, especially in the environmental area. The contribution of the remaining sectors, for example, the manufacturing and market services sector, to GVA remained more or less constant throughout the whole period analysed.

Consumption of fuels used for heat and electricity generation or for technological processes was also stable over the whole period. During 2000–2004, fuel

Table 11.2 Contribution to GVA by nine branch categories (%)

Branch	*Section*		*1997*	*2000*	*2004*	*2007*
nace9–1	A,B	Agriculture	4.4	4.1	3.5	2.6
nace9–2	C	Mining	1.9	1.5	1.4	1.3
nace9–3	D	Manufacturing	28.6	28.1	28.2	28.1
nace9–4	E	Power	4.1	3.5	4.1	4.5
nace9–5	F	Construction	7.9	6.8	6.8	6.8
nace9–6	G	Trade	12.1	14.5	12.1	13.4
nace9–7	I	Transport	7.9	6.9	7.8	7.8
nace9–8	H,J,K	Market services	16.7	17.9	17.9	17.7
nace9–9	L,M,N,O	Public services	16.5	16.8	18.1	17.7

consumption declined continuously but overall total use of fuels declined by only less than 2 per cent in total. The only large decline in fuel use occurred in 1999, as an effect of economic recession.

Air emissions released by stationary emission sources dropped significantly. During the 11 years analysed, emissions of CO declined by 67 per cent, SO_2 by 72 per cent and PM by 89 per cent. Most of the reductions occurred up until 2000 as a result of the Air Quality Act No. 389/1991, which required stationary emission sources (such as power plants and industrial processes) to meet a strict emission limit prior to 1998.[16] Máca *et al.* (2010) found that although the rate of emission charges was continuously increasing nominally, this instrument was neither effective nor efficient in combating emissions. The command-and-control regulation was in fact the only effective instrument in the Czech Republic to regulate emissions from stationary sources. Since 2000, emissions of particulate matters have mainly been reduced further by about 12 per cent during each of the two periods, 2000–2004 and 2004–2007. Emissions of CO followed the trend in reductions in emissions of PM.

We observe a quite different situation for emissions of NO_x. At first these emissions were reduced by almost 18 per cent during the period 1997–2000 due to an increase in technological efficiency. In the next period, 2000–2004, emissions of NO_x from stationary sources increased (by 4.6 per cent) and in the later period decreased slightly (by 2 per cent in total).

Decomposition results

The four-factor decomposition analysis

Results of our decomposition analysis undertaken for changes in four air pollutants, specifically SO_2, PM, CO and NO_x, are displayed in Appendix (Figures 11.A.1 to 11.A.4). First, we can discern a similar trend in the driving-force effects in question on emissions of PM and SO_2 whereas the drivers play a rather different role both in terms of direction and magnitude in affecting emissions of NO_x.

A change in the economic structure as captured by the composition factor has both positive and negative effects on emission changes. The composition effect is negative particularly during the period 1997–2000 and for SO_2 and NO_x emissions. Moreover, in most cases, it even largely exceeds the effect of scale. During this period, economic production declines particularly in the agriculture, power, transport and mining sectors which all belong to relatively emission-intensive sectors.

A change in the economic structure leads to increases in pollution, especially in two years, 2004 and 2006, when the Czech economy was also significantly increasing in scale. In both of these years, we also document a significant increase in the share of power production that usually belongs to more emission-intensive sectors. However, we would like to stress here that due to completion of the second Czech nuclear power plant (Temelín), electricity generation from nuclear power has doubled since 2003.[17] As a result, the share of nuclear electricity increases from 20 to 30 per cent of total electricity production and the overall increase in electricity production (by 10 per cent in 2003 compared to 2000) is associated with fewer air emissions because of the zero air emissions of nuclear power production. We therefore assume that the positive effect of the composition factor in 2004 and 2006 is probably due to structural changes within the manufacturing sector.

One reason for the existence of the environmental Kuznets curve is that as the economy develops, its structure moves from 'dirtier' to 'cleaner' production. We can confirm this type of trajectory during the end period of the economic and political transformation in the Czech Republic, up until 2000, although since then the change in economic structure has been more pollution-intensive rather than resource-saving.

In general, there are four emission abatement options. With the exception of reducing output, emissions can be abated either by the installation of end-of-pipe technology (using a filter), changing a fuel mix or improving technology/production efficiency that results in using less fuel per unit of product.

The last option is captured by the *fuel-intensity effect* in our decomposition analysis. Looking at aggregate statistics, we find that, excluding the year 2001, aggregate fuel intensity as measured by GJ of fuel used per unit of GVA produced by all stationary emission sources is declining over time. However, fuel intensity in each sector might in reality decline or increase. The observation based on macro statistics is confirmed by our decomposition analysis which shows that the fuel-intensity effect is negative (resulting in emission reduction) in most of the years, but mainly in the period from 2002 to 2006. We find the strongest reduction effect in fuel intensity in 2006 (about 20 per cent), and then in 2002 and 2004 when it is about 7 to 9 per cent, whereas it is less than 5 per cent in the remaining years in that period. We observe a negative effect of fuel intensity (more or less the same in magnitude) for all four analysed pollutants. Interestingly enough, fuel intensity leads to a larger reduction in emissions than in the emission–fuel coefficient does during the entire period. Some emission reduction can also be attributed to fuel intensity during the period 1997–1999 although its effect is quite small in absolute magnitude and very small compared with the effect related to the emission–fuel coefficient factor.

Indeed, the emission–fuel coefficient was the strongest and most dominant factor to abate air emissions within stationary emission sources in the Czech Republic until 1999 (SO_2 and NO_x) or until 2000 (PM and CO), respectively. As we argued earlier, this factor captures abatement through the installation of end-of-technologies and through a fuel switch towards using environmentally friendly energy carriers.

The emission–fuel coefficient reflects the enormous abatement effort of Czech enterprises during the 1990s that accompanied large investment activity. In fact, once the Act on Air Quality was announced in 1991, regulated firms started to invest in abatement technologies in order to comply with the emission standard required by the respective law by the year 1998. The abatement effort of firms can be measured by environmental investment spent on air quality and climate change protection. Looking at the national statistics, we can indeed observe a peak in investment activity in 1996 when a total of almost CZK 27 billion (2000 prices) was spent. Abatement activity remains quite large in later years when air-quality-specific investment reached about 77 per cent and 59 per cent of the 1996 top level in 1998 and in 1999, respectively (see Air-Inv in Figure 11.2). From 2000 onwards, after reaching the emission target, the abatement of firms relaxes and since 2002, investment has stabilized at about 13 to 15 per cent of the 1996 level. The slow-down in the abatement effort is also reflected in the smaller effect of the emission–fuel factor in later years.

Such a distinction between the fuel-intensity and emission–fuel coefficient effect cannot be made in a three-factor decomposition. The reason is that the *emission-intensity factor* in the three-factor analysis captures all three abatement channels (i.e. end-of-pipe technology, fuel switch and increasing technology/production efficiency). It can therefore only indicate the effect of the overall environmental friendliness of production, but cannot attribute emission reduction to any specific channel in the same way as the four-factor analysis can. We document the value added of the four-factor analysis in Figure 11.3. It shows the emission–fuel intensity (dark grey) and fuel-intensity effect (light grey) together with the intensity effect (black) which actually aggregates the previous two effects into one. In the years when one (emission–fuel coefficient) factor dominates, the difference between a three- and four-factor decomposition analysis is not great, but in later years, the four-factor analysis with more refined decomposition provides more information on the emission reduction driver.

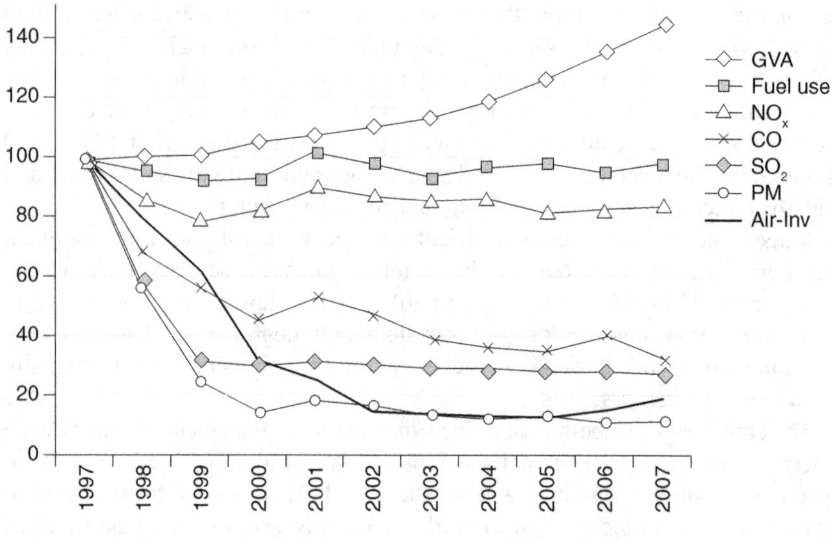

Figure 11.2 Overall trend in economic performance, air emissions and fuel consumption.

Note

Air-Inv denotes investment in air quality and climate change when its 1997 level is set to 100.

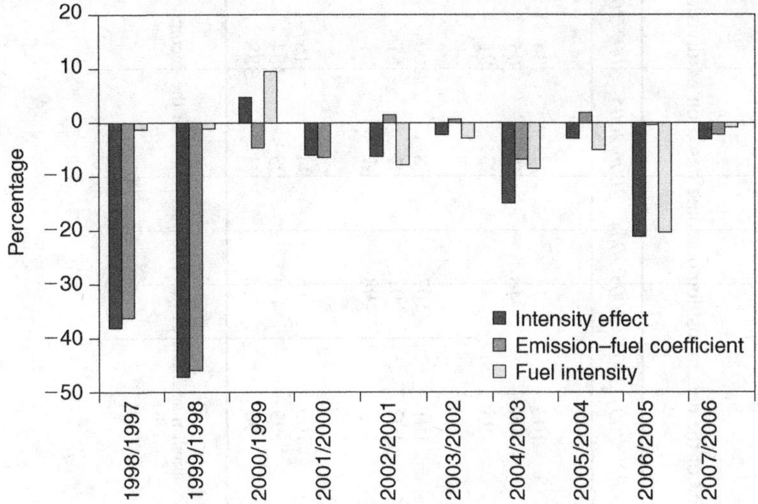

Figure 11.3 Decomposition of the intensity effect into the emission–fuel coefficient and fuel intensity, changes in SO_2 emissions 1997–2007.

Decomposition for various sector breakdowns

As we discussed earlier, the decomposition results differ when the decomposition analysis is based on differently disaggregated data. In general, as shown by Rørmose and Olsen (2003), the more disaggregated the driving forces are, the larger the effects identified by the decomposition analysis. Moreover, as shown by Seibel (2003), reducing the level of sector breakdown might lead to a biased result for the intensity and structure effects in particular.

To find the effect of different sector breakdowns, we undertook our decomposition analysis for four various levels of sector breakdown. Our base case is based on the two-digit codes of the NACE nomenclature (with 54 divisions of sectors in the Czech economy). We then decomposed changes in air emissions with the economy broken down into 30 sub-sections (based on two-character alphabetical codes), 16 sections (letters A to Q) and lastly for nine wider categories. Table 11.3 displays the result for PM emission decomposition for each level of sector breakdown compared to the result based on our finest sector disaggregation based on 54 divisions (set to 100).[18] We summarize our result (in the same way as Seibel, 2003 applied SDA to CO_2 emissions in Germany) so that the less disaggregated the economy, the larger the bias – especially, the composition and emission–fuel coefficient. Specifically, in most years the composition effect for a less detailed sector breakdown gets smaller. It basically indicates that part of emission changes can be attributable to structural changes within groups of branches especially if we break down the economy into subsections or even sections.

Table 11.3 Decomposition of PM emissions for various levels of sector break-down (magnitude of the decomposition outcome based on 54 divisions set to 100)

	1998/1997	*1999/1998*	*2000/1999*	*2001/2000*	*2002/2001*	*2003/2002*	*2004/2003*	*2005/2004*	*2006/2005*	*2007/2006*
Composition effect										
Subsections (30)	91	83	86	95	–17,396	142	102	73	74	134
Sections (16)	64	8	95	99	–26,092	173	98	40	83	95
Groups (9)	64	9	97	99	–27,347	170	98	42	82	91
Emission–fuel coefficient										
Subsections (30)	98	102	103	123	92	103	101	107	58	100
Sections (16)	103	100	105	152	185	124	102	91	101	81
Groups (9)	103	100	104	148	190	124	102	98	87	81
Fuel intensity										
Subsections (30)	127	367	74	73	109	86	100	16	94	–16
Sections (16)	96	–194	6	30	94	56	95	–491	93	384
Groups (9)	96	–187	15	35	93	56	96	–364	94	388

Note
The extremely large differences of the composition effect during the period 2002/2001 are due to the very small magnitude of this effect as found for the decomposition analysis based on 54 divisions.

For our example, we conclude that if we base our analysis on a less detailed sector breakdown, especially at a lower than sub-section level, then we might overstate the effect of emission–fuel coefficient and understate the effect of change in composition. The mistake is particularly serious during the period when emissions decrease overall. In some cases, we could even provide a wrong result by indicating the negative effect of fuel intensity that was actually positive. Our result therefore supports applying as detailed a sector disaggregation as possible in an index decomposition analysis.

Overall, during the whole period 1997–2007, the change in economic structure is PM-, SO_2-, and NO_x-reducing. There are, however, some periods with a positive effect of structural change in the economy on emissions, particularly during the later period 2004 on PM and SO_2 and during early period 1997–2004 on NO_x; all of these effects were however small in magnitude in the range of 2 to 4.5 percent.

Decomposition for various time aggregations

We decompose the change in air emissions for three distinct periods. Specifically, the first period 1997–2000 can be characterized by the large abatement effort of Czech firms to meet the strict requirement to significantly reduce their air emissions. The second period, 2000–2004, is characterized by company and governmental efforts to comply with the legal requirements of the EU *acquis communitaire* by *adjusting* institutions to the EU standard and the trade patterns of companies in order to become competitive in the European single market. The last period, 2004–2007, is the period when the Czech Republic is already a full member of the European Union. Figure 11.4 then reports on the aggregated result of our decomposition for the entire period 1997–2007 and for the three consequent sub-periods. The contribution of each of the four determinants of air emissions for each period is computed from the year-by-year based decompositions (i.e. we aggregate the effect across each year of the period). The total change in emission level during the given period is displayed on the right of each figure.

As we can see, most emission abatement occurs in the first period up until 2000. However, PM emissions are reduced in the next two periods by 12 per cent and we point out that the abatement effort reduces emissions from a much smaller level due to the enormous reduction up until 2000. In total, emissions of PM are only reduced by 3.4 per cent during the period 2000–2007 compared with the initial 1997 level. After 2000, we also find a slow-down in abatement activity for other pollutants. During the period 2004–2007, total emissions of SO_2 and NO_x are reduced slightly, by less than 1 per cent and 2 per cent respectively, while emissions of NO_x increase by 4.6 per cent in the period 2000–2004.

The change in fuel intensity is emission-reducing for PM and SO_2, and slightly emission-increasing for NO_x throughout the entire period 1997–2007. Interestingly enough, the fuel-intensity effect plays a different role in changes in emissions of NO_x at the beginning and the end of the period; it is particularly strong and positive during the period 2004–2007 when the Czech economy grew at the largest rate in its history, whereas it is the second strongest driver to reduce these emissions from 1997 to 2000. For NO_x at least, our result indicates that technology changes introduced during the economic recession in 1997–1999 were pollution-saving while changes in technology during the economic boom that occurred in the Czech Republic during 2004–2007 were rather pollution-using.

The emission–fuel coefficient is always negative which reflects the effect of using clean end-of-pipe technologies, except in the 2000–2004 period when the effect is slightly positive for SO_2 emissions.

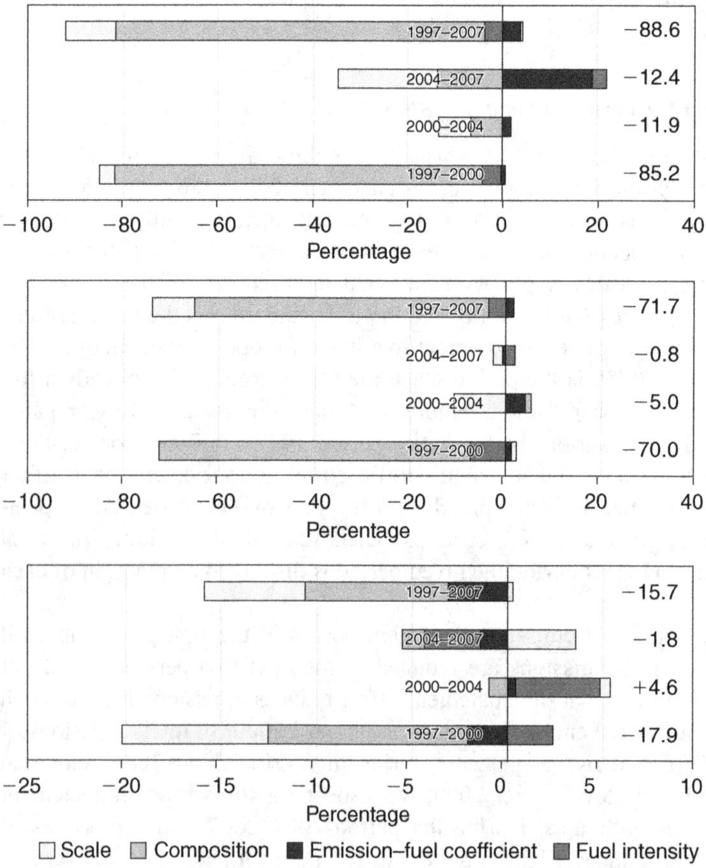

Figure 11.4 Decomposition of change in PM, SO_2 and NO_x emissions for several periods, aggregated from year-by-year decompositions.

Note

The residuals are 5 to 6 order of magnitude smaller than the total change. Due to their very small magnitude, we have not shown the residual term in the figure.

Lastly, as argued by Löfgren and Muller (2010), decomposition based solely on the first and final year in the period might result in a biased result, and, if possible, decomposition analysis should be performed on the year-by-year changes. This is exactly the approach we have applied so far. How the result can be biased by undertaking decomposition analysis on a period-wise data is documented in Table 11.4. It reports our result for the decomposition of PM emissions as undertaken on year-by-year as well as period-wise data. We also compute the difference between both results and express them as a percentage deviation of the period-wise approach from the more correct result based on year-by-year analysis. This difference informs us of the error of decomposition analysis when we only rely on changes in the first and end year of the period. As we can see, the decomposition based on a period-wise analysis generates an erroneous result not only in the effect's magnitude, but also, in some cases, in the effect's direction. These differences appear even for quite short periods.

Conclusions

A relationship between economic performance and environmental burden is often analysed through testing the environmental Kuznets curve hypothesis. These studies usually rely on a reduced form econometric analysis of estimating the effect of per capita income on environmental degradation levels. The results of reduced form analysis do not necessarily lead to clear policy recommendations. On the other hand, decomposition analysis can identify through which channel environmental burden is affected by economic variables and thus provide clearer policy recommendation.

In our study, we utilize an index-based statistical decomposition to examine which factors were active in changing the emission level of four pollutants – SO_2, NO_x, CO and particulate matters – during the transition and post-transition period of the Czech economy from 1997 to 2007. Unlike the decompositions based on the Laspeyres index, which generate quite large residuals especially when there are substantial changes (which is actually the Czech situation in the analysed period), we employ a method based on the Divisia index to overcome this problem. Following Löfgren and Muller (2010), we consider annual changes rather than basing our decomposition analysis on longer time intervals to avoid the biased results.

In our four-factor decomposition analysis we find that the change in economic structure leads to both positive and negative effects on changes in emissions in question. The composition effect is negative particularly during the period 1997–2000 and for SO_2 and NO_x emissions, and in most cases even exceeds the effect of scale. Change in economic structure leads to increases in pollution especially in the two years, 2004 and 2006, when the Czech economy was also significantly increasing in scale. Our more detailed examination leads us to conclude that this positive effect is probably due to structural changes within the manufacturing sector.

Table 11.4 Decomposition of PM for the period-wise and year-by-year approach (%)

	1997–2000			*2000–2004*			*2004–2007*		
	Initial-final period	*Year-by-year*	*Difference*	*Initial-final period*	*Year-by-year*	*Difference*	*Initial-final period*	*Year-by-year*	*Difference*
Scale	4.5	0.7	529	12.5	1.9	571	20.8	18.6	12
Composition	–4.7	–4.3	9	8.1	–0.7	–1,262	4.3	3.7	13
Emission–fuel coefficient	–81.7	–77.3	6	–14.5	–6.6	121	–14.1	–14.4	–2
Fuel intensity	–3.2	–4.3	–24	–17.9	–6.5	176	–23.4	–20.3	15
Residuals	0.0	0.0	–5	0.0	0.0	–334	0.0	0.0	–47
Total change	**–85.2**	**–85.2**		**–11.9**	**–11.9**		**–12.4**	**–12.4**	

The emission–fuel coefficient was the strongest and most dominant factor to abate air emissions stemming from stationary emission sources in the Czech Republic until 1999/2000. In our study, we are therefore able to distinguish the abatement effect due to output reduction, the one due to fuel intensity changes and lastly, the effect of abatement reached through the installation of end-of-technologies or through a fuel switch towards using environmentally friendly energy carriers. We note that such a distinction particularly between the fuel-intensity and emission–fuel coefficient effect cannot be made in a three-factor decomposition which can only indicate the effect of the overall environmental friendliness of production, but cannot attribute emission reduction to any specific channel in the same way as four-factor analysis.

To find the effect of different sector breakdowns, we undertook our decomposition analysis for four various levels of sector breakdown. Our base case is based on the two-digit codes of the NACE nomenclature (54 divisions of sectors in the Czech economy). Then we decomposed changes in air emissions with the economy broken down into 30 sub-sections, 16 sections and lastly, nine wider categories. We summarize our result so that the less disaggregated the economy, the larger the bias. Specifically, in most years the composition effect for a less detailed sector breakdown gets smaller. This basically indicates that part of emission changes can be attributed to structural changes within groups of branches. Our result therefore supports applying as detailed a sector disaggregation as possible in decomposition analysis.

After aggregating year-by-year changes over the whole period 1997–2007, we find that the change in economic structure is PM-, SO_2- and NO_x-reducing. Overall, the change in fuel intensity is emission-reducing for PM and SO_2, and slightly emission-increasing for NO_x during the entire period 1997–2007. Interestingly enough, the fuel-intensity effect plays a different role in changes in emissions of NO_x at the beginning and the end of the analysed period; our result specifically indicates that technology changes introduced during the economic recession in 1997–1999 were pollution-saving whereas changes in technology during the economic boom during 2004–2007 were rather pollution-using. The emission–fuel coefficient is always negative which reflects the effect of using clean end-of-pipe technologies, except for the period 2000–2004 when the effect is slightly positive for SO_2 emissions.

Lastly, as argued by Löfgren and Muller (2010), decomposition based solely on the first and final year in the period might result in a biased result, and decomposition analysis should be performed on the year-by-year changes. Our study also makes this recommendation. How large this bias might be we document by computing a difference between decomposition results based on the period-wise approach and on year-by-year analysis.

Appendix

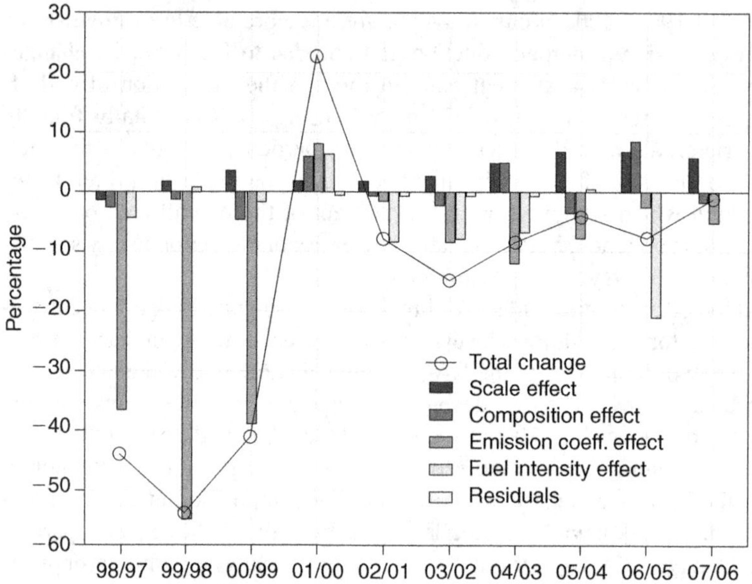

Figure 11.A.1 Additive particulate matter decomposition – NACE54.

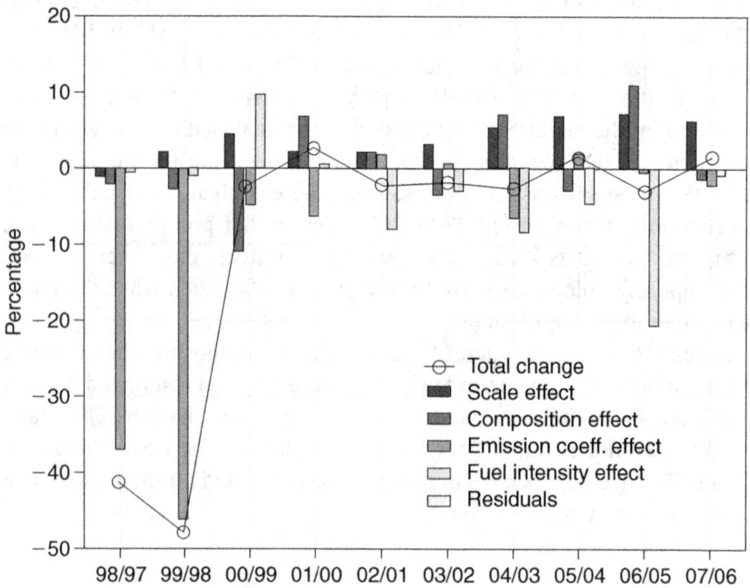

Figure 11.A.2 Additive SO_2 decomposition.

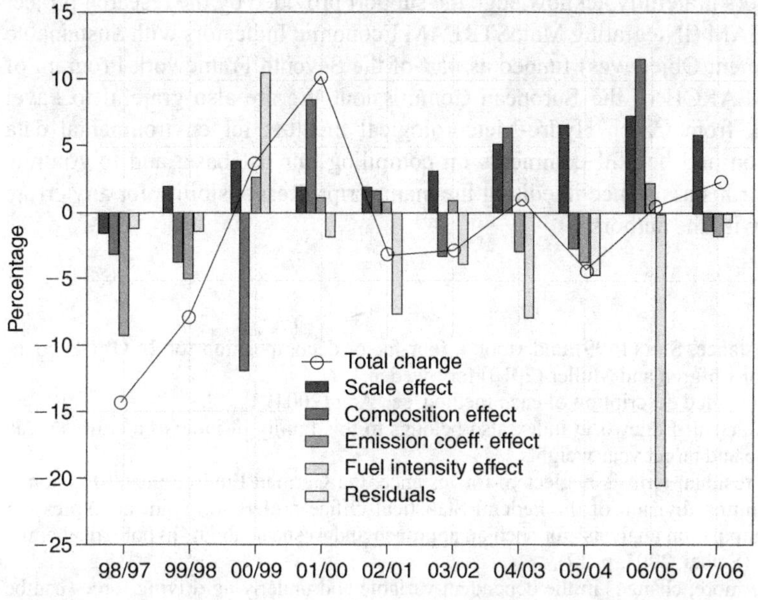

Figure 11.A.3 Additive NO_x decomposition – NACE54.

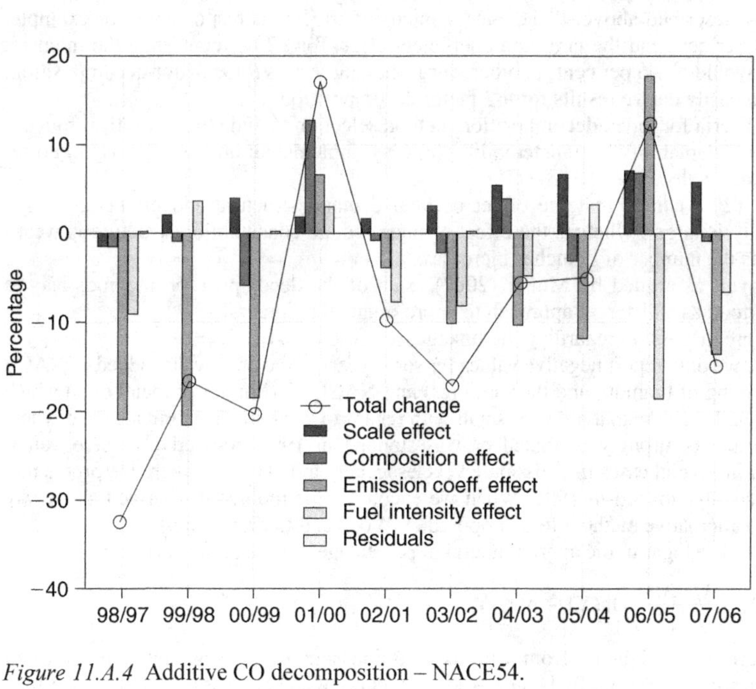

Figure 11.A.4 Additive CO decomposition – NACE54.

Acknowledgements

The authors gratefully acknowledge the support provided by the research project IN-STREAM (INtegrating MainSTREAM Economic Indicators with Sustainable Development Objectives) funded as part of the Seventh Framework Program of DG RESEARCH of the European Commission. We are also grateful to Pavel Machálek from Czech Hydro-Meteorological Institute for environmental data preparation and helpful comments on compiling our database, and to Andrew Barton for his assistance in editing the manuscript. Responsibility for any errors remains with the authors.

Notes

1 For instance, Sun (1999) undertook a four-factor decomposition for 24 OECD countries, or Löfgren and Muller (2010) for Sweden.
2 For a detailed description of each method, see Ang (2004).
3 The Marshall-Edgeworth index also belongs to this family of indexes and uses mean of base and target year weights.
4 If the residual term is neglected, for instance, the German Environmental Economic Accounting division of the Federal Statistical Office prefers to avoid the expression 'decomposition analysis' for such an approach and to speak about 'hypothetical emissions' (Seibel, 2003, p. 17).
5 Furthermore, changes in the dependent variable and underlying driving forces can be expressed in absolute terms or relative terms.
6 Imagine, for instance, that we decompose change in emission into four factors as already described above. If the energy-intensity effect was responsible, for example, for –8 per cent and the emission-coefficient effect for +2 per cent, then the intensity effect would be –6 per cent. In other words, having results for *n*-factor decomposition, we can easily derive results for *n-1* factor decomposition.
7 Four criteria for index decomposition method selection include (a) theoretical foundation, (b) adaptability or transferability, (c) easy implementation, and (d) easy interpretation of results.
8 Seibel (2003) finds that the effect of final demand structure and effect of energy intensity is larger, whereas the effect of intermediate consumption structure is weakened as the number of branches increases.
9 Moreover, as argued by Muller (2007), each of the decomposition methods can be understood as a different approach to approximate that integral.
10 For more discussion regarding the linkage, see Ang (2004).
11 In fact, we only report negative values for some years in the gross value added of NACE 12 (Mining of uranium and thorium ores) and NACE 13 (Mining of metal ores) which are actually sectors that are very small with regard to economic performance and where production is strongly regulated or was stopped in 1995, respectively. Zero values appear in several cases in emission levels, especially in NACE 13 where the production was actually stopped in 1995. When we encounter the problems described above, we use the alternative methodology proposed by Wood and Lenzen (2006).
12 Here we use logarithmic approximation of percentage change of the variable:

$$\Delta X_t = \frac{X_t - X_{t-1}}{X_t} \approx \ln(X_t) - \ln(X_{t-1})$$

13 Air emissions might be from various emission sources such as stationary, area or mobile sources or from fuel combustion or technological processes without using fuel. A change in the composition of the emission sources and/or type of process is not,

however, included in our composition effect. We strictly keep the effect to capture the change in economic structure.

14 Tsuchimoto and Ščasný (2011) introduce the *fuel-mix factor* in their five-factor analysis to decompose air emission in the Czech Republic. Their analysis thus distinguishes the effect of abatement technology, which is usually represented by end-of-pipe type technology, from fuel switches.

15 Definition of nine branch categories is based on 17 branch sections (defined by letters A to Q). We merge section A and B into one, sections H, J and K into another one, and sections L, M, N, O, P into the next one. Section Q is not included in our analysis.

16 The emission limit was further specified by decree 117/1997, which came into force on 1 June 1997 and was set by the Czech Environmental Inspection by 31 December 1998. Medium-sized emission sources were obliged to measure their emission releases and thus meet the specified emission target from 1998 onwards.

17 Nuclear power has increased electricity production from about 13 TWh during the 1990s up to about 26 TWh since 2003. Since that year, the Czech Republic has also become one of the largest exporters of electricity both in Europe and worldwide, with almost 20 per cent of exported electricity production.

18 We draw a similar conclusion when we decompose SO_2 and NO_x emissions with a different sector disaggregation as we did for PM emissions. These results are available upon request.

References

Ang, B.W. (1995) 'Decomposition methodology in industrial energy demand analysis', *Energy*, 20: 1081–1095.

Ang, B.W. (2004) 'Decomposition analysis for policymaking in Energy: which is the preferred method', *Energy Policy*, 32: 1131–1139.

Ang, B.W. (2006) 'Monitoring changes in economy-wide energy efficiency: from energy GDP ratio to composite efficiency index', *Energy Policy*, 34: 574–582.

Ang, B.W. and Liu, N. (2001) 'A new energy decomposition method: perfect in decomposition and consistent in aggregation', *Energy*, 26: 537–548.

Brůha, J. and Ščasný, M. (2006) 'Economic analysis of driving forces of environmental burden during the transition process: EKC hypothesis testing in the Czech Republic', paper presented at the 3rd Annual Congress of Association of Environmental and Resource Economics AERE, Kyoto, 4–7 July.

Cherp, A., Kopteva, I. and Mnatsakanian, R. (2003) 'Economic transition and environmental sustainability: effects of economic restructuring on air pollution in the Russian Federation', *Journal of Environmental Management*, 68: 141–151.

Diekmann, J., Eichhammer, W., Neubert, A., Rieke, H., Schlomann, B. and Ziesing, H.J. (1999) 'Energie-Effizienz-Indikatoren. Statistische Grundlagen, theoretische Fundierung und Orientierungsbasis für die politische Praxis (Umwelt und Ökonomie Bd. 32) (Umwelt Und Okonomie)', Heidelberg.

Hoekstra, R. and van den Berg, J.C.J.M. (2003) 'Comparing structural and index decomposition analysis', *Energy Economics*, 25: 39–64.

Jungnitz, A. (2008) 'Decomposition analysis of greenhouse gas emissions and energy and material inputs in Germany', paper prepared within Petre Project – Resource Productivity, Environmental Tax Reform and Sustainable Growth in Europe, mimeo. Online: www.petre.org.uk/papers.htm (accessed 5 February 2011).

Liu, N. and Ang, B.W. (2007) 'Factors shaping aggregate energy intensity trend for industry: energy intensity versus product mix', *Energy Economics*, 29: 609–635.

Löfgren, A. and Muller, A. (2010) 'Swedish CO_2 emission 1993–2006: an application of decomposition analysis and some methodological insights', *Environmental and Resource Economics*, 47: 221–239.

Máca, V., Melichar, J. and Ščasný, M. (2010) 'External costs from energy generation and their internalisation in new member states', in C.D. Soares, J. Milne, H. Ashiabor, K. Deketelaere and L. Kreiser (eds), *Critical Issues in Environmental Taxation*, Vol. VIII, Oxford: Oxford University Press.

Muller, A. (2007) 'Putting decomposition of energy use and pollution on a firm footing: clarifications on the residual, zero and negative values and strategies to assess the performance of decomposition methods', *Scandinavian Working Papers in Economics*, 215.

Muller, A. (2008) 'Clarifying poverty decomposition', *Scandinavian Working Papers in Economics*, 217.

Nanduri, M. (1996) 'An assessment of energy intensity indicators and their role as policy-making tools', mimeo, Vancouver.

Ritschelová, I., Machálek, P. and Koroluk, R. (2001) 'NAMEA for the Czech Republic 1992–1996', in *NAMEAs for Air Emissions (Results of Pilot Studies)*, Luxembourg: European Commission.

Rørmose, J.P. and Olsen, T. (2003) 'Analysis of changes in air emissions in Denmark 1980–2001: time series', Bridge tables, Decomposition analysis, mimeo, Copenhagen.

Seibel, S. (2003) 'Decomposition analysis of carbon dioxide-emission changes in Germany: conceptual framework and empirical results', mimeo, Luxemburg.

Sun, J.W. (1998) 'Changes in energy consumption and energy intensity: a complete decomposition model', *Energy Economics*, 20: 85–100.

Sun, J.W. (1999) 'The nature of CO_2 emission Kuznets curve', *Energy Policy*, 27: 691–694.

Tol, R.S.J. and Weyant, J.P. (2006) 'Energy economics' most influential papers', *Energy Economics*, 28: 405–409.

Tsuchimoto, F. and Ščasný, M. (2011) 'Decomposition analysis of air pollutant in the Czech Republic', paper submitted to 18th Annual Conference of the European Association of Environmental And Resource Economists (EAERE), Rome, 29 June–2 July.

Vehmas, J., Luukkanen, J. and Pihlajamäki, M. (2008) 'Decomposition analysis of EU sustainable development indicators', Final Report of the DECOIN project (Development and Comparison of Sustainability Indicators), Deliverable D3.1 of WP3.

Viguier, L. (1999) 'Emissions of SO_2, NO_x and CO_2 in transition economies: emission inventories and Divisia Index analysis', *Energy Journal*, 20: 59–87.

Wood, R. and Lenzen, M. (2006) 'Zero-value problems of the logarithmic mean Divisia Index de-composition method', *Energy Policy*, 34: 1326–1331.

Index

Page numbers in *italics* denote tables.

9781138807600